ROUTLEDGE LIBRARY EDITIONS:
AGRICULTURE

Volume 12

AN EMPIRICAL INVESTIGATION OF FARMERS' BEHAVIOR UNDER UNCERTAINTY

AN EMPIRICAL INVESTIGATION OF FARMERS' BEHAVIOR UNDER UNCERTAINTY

Income, Price and Yield Variability for
Late-Nineteenth Century American Agriculture

ROBERT A. MCGUIRE

Routledge
Taylor & Francis Group

LONDON AND NEW YORK

First published in 1985 by Garland Publishing, Inc.

This edition first published in 2020
by Routledge
2 Park Square, Milton Park, Abingdon, Oxon OX14 4RN

and by Routledge
52 Vanderbilt Avenue, New York, NY 10017

Routledge is an imprint of the Taylor & Francis Group, an informa business

British Library Cataloguing in Publication Data
A catalogue record for this book is available from the British Library

ISBN: 978-0-367-24917-5 (Set)
ISBN: 978-0-429-32954-8 (Set) (ebk)
ISBN: 978-0-367-25205-2 (Volume 12) (hbk)
ISBN: 978-0-429-28652-0 (Volume 12) (ebk)

Publisher's Note
The publisher has gone to great lengths to ensure the quality of this reprint but points out that some imperfections in the original copies may be apparent.

Disclaimer
The publisher has made every effort to trace copyright holders and would welcome correspondence from those they have been unable to trace.

AN EMPIRICAL INVESTIGATION
OF FARMERS' BEHAVIOR
UNDER UNCERTAINTY ★ Income, Price and
Yield Variability for
Late-Nineteenth Century
American Agriculture

Robert A. McGuire

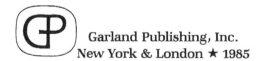

Garland Publishing, Inc.
New York & London ★ 1985

Library of Congress Cataloging-in-Publication Data

McGuire, Robert A. (Robert Allen), 1948–
  An empirical investigation of farmers' behavior under
uncertainty.

  (American economic history)
  Thesis (Ph. D.)—University of Washington, 1978.
  Bibliography: p.
  1. Agriculture—Economic aspects—United States—
History—19th century. I. Title. II. Series:
American economic history (Garland Publishing, Inc.)
  HD1761.M395 1985   338.1'0973   84-48311
  ISBN 0-8240-6659-6

All volumes in this series are printed on acid-free,
250-year-life paper.

Printed in the United States of America

PREFACE

When I decided to publish this dissertation and was given the
opportunity to make revisions, my first thought was to update the entire
study. Upon reflection, I realized that a complete revision would likely
lead the dissertation from its course and into areas not originally
contemplated. I concluded that this type of revision should be
appropriately left for future study. Additionally, because several parts
of the study have been published in revised form as journal articles, the
need for certain revisions has diminished. For these reasons, I have
decided against making any revisions and have published the dissertation as
originally written.

I think it would be useful to readers, nevertheless, to be aware of
the changes that I might have made--if I were to have made any. First, a
more detailed explanation of the limitations of my random price, yield, and
income variability estimates (Chapter III) could have been offered. These
estimates are computed for the purpose of rigorously measuring the risks
associated with farming for all major crops in the United States during the
second half of the nineteenth century. Because my concern is with all
agriculture during this period, statewide data are used for these
estimates. The use of these data does create some problems of aggregation.
Ideally, a large sample of individual farm yields and prices would be used.
However, individual farm data for all states during the period under study
do not exist.

Price variability computed from these data (with trend removed
through the variate difference method) will not significantly misrepresent

the variability faced by individual farmers. The individual farmer virtually had no control over the price received nor over the degree of yearly price variability, because of the competitiveness of the determination of agricultural prices. The individual farmer, therefore, faced a common market price and its fluctuations, which as a first approximation can be estimated accurately with statewide data.

The case of yield variability is different, however. Ideally, yield variability would be based on a sample of yields occurring under constant technology, prices, and costs, but subject only to random disturbances--for example, weather. Such a data set does not exist. Estimates of yield variability where factors which could affect production decisions are not held constant--but where trend is removed--would seriously misrepresent a "true" measure of risk only to the extent that year-to-year changes in managerial decisions--for example, changes in input levels--affect yields in a random manner. Detrending should minimize the impact on yield fluctuations of most production decisions, which would typically affect yields in a systematic or nonrandom manner.

To the extent that some managerial decisions affect yields randomly, my estimates of yield variability overstate the "true" amount of risk faced by individual farmers. On the other hand, to the extent that statewide yields "average out" individual farm differences in variability, the variability estimates presented in this study understate yield variabilities faced by individual farmers. Given that my interest is in the relative nature of yield variabilities (not the exact magnitude) among different crops and across states, the estimates computed from statewide data are expected to reasonably indicate such significant differences.

This follows because production decisions, which create random fluctuations in yields, would not be expected to vary systematically among different crops or across states.

Because my estimate of gross income per acre--an approximation of returns per acre--is the product of price and yield, my estimate of its variability using statewide data is no more limited than the price and yield variability estimates. This leads to the second change I might have made. Because the year-to-year correlation between prices and yields for each crop also is important in determining the level of income variability, I probably should have computed my own estimates of the correlation between the year-to-year changes in prices and yields. Instead I relied on correlation estimates contained in the existing literature. Since the dissertation's completion, I have computed my own correlation estimates. They provide even stronger support than the estimates used in the dissertation for all statements contained Chapter III about the influence of yearly price-yield correlation on income variability.

The third change I might have made also would have been in Chapter III. I might not have included the estimates of the variability of the quantity and value per head of livestock. These estimates may not be very accurate measures of the risks associated with raising livestock. The problem mainly concerns the variability estimates of livestock quantities. These estimates may have captured primarily the effects of farmers' output decisions, not unpredictable or random changes in births and deaths due to unforeseen circumstances. Only the random variabilities would be measures of the risks of livestock farming. If these quantity changes primarily reflect farmers' output decisions, the value per head variabilities are also less than accurate. In this case, much of my estimated price (value per head) variabilities may be including the influence of these decisions. As with the yield variability estimates, however, this problem with the livestock variability estimates, where trend is removed, is more serious only to the extent that year-to-year changes in managerial decisions affect livestock quantities (and, thus, prices) in an unpredictable or random manner.

The final change I might have made concerns Chapter V. In this chapter, I use my random variability estimates to explain the location and timing of the various agrarian protest movements of the last third of the nineteenth century. I argue that these movements were a response to the high levels of price, yield, and income variability faced by farmers. While the cross-sectional analysis provided strong support for an explanation of the location of these protest movements, the time-series analysis provided little explanation for their timing. Instead of relying on the single-causal tests for the time-series analysis, I probably should have employed regression analysis to capture the obvious multi-causal nature of the problem, where the riskiness in farming is only one of several factors.

I also would like to use this opportunity to inform readers where they can locate the parts of this study that have been published in revised form. The contents of Chapter IV have been published as "A Portfolio Analysis of Crop Diversification and Risk in the Cotton South" in Explorations in Economic History (October 1980). This article contains more detail than Chapter IV on the controversy between myself and Gavin

Wright and Howard Kunreuther about the output decisions of postbellum farmers in the face of agricultural instability.

The cross-sectional analysis of agrarian unrest contained in Chapter V has been published as "Economic Causes of Late-Nineteenth Century Agrarian Unrest: New Evidence" in the Journal of Economic History (December 1981). This article presents even stronger statistical evidence than Chapter V that the location of the agrarian protest movements can be explained by the level of agricultural instability. My response to a comment by Bradley Lewis that primarily is an expansion of my 1981 paper has been published as "Economic Causes of Late-Nineteenth Century Agrarian Unrest: Reply" in the Journal of Economic History (September 1982). This reply contains both a discussion of Lewis' weaknesses and my own corrections to a statistical error in my 1981 paper.

Finally, an updated version of the introduction to Appendix A has been published as "U.S. Agricultural Statistics: State Estimates, 1866-1914" in Agricultural History (April 1980). This article presents a complete listing of all crop and livestock data sources provided by the U.S.D.A. on a continuous annual statewide basis for 1866-1914. It also contains a brief discussion of the statewide data and a listing of those series which are consistently incomplete for all states or several years. A full list of missing observations for the eleven principal crops and six species of livestock for each state has been published as "A Complete List of Missing Observations for U.S.D.A. Statewide Data, 1866-1914" in Bureau of Business Research, Ball State University, Working Paper No. 13 (January 1981).

Robert A. McGuire

Muncie, Indiana
March 1985

To my parents,

      Russell and Viola,

with love

ACKNOWLEDGMENTS

I would like to thank my doctoral committee members for their assistance in the preparation of this dissertation. They are Professors Vernon Carstensen, Richard Hartman, Robert Higgs, and Douglass C. North. I am especially thankful to Professor Robert Higgs, my doctoral committee chairperson, whose assistance far exceeded any requirement. He has been extremely helpful with his encouragement and quite liberal with his time.

I also thank Professor Potluri Rao of the Department of Economics of the University of Washington and Dr. Josef Schwalbe of ABT Associates Inc. for their programming assistance.

Ms. Kathleen Kurz typed the drafts of the dissertation, while Mrs. Barbara Miller did the final manuscript.

Finally, I am indebted to my parents, Russell and Viola McGuire, who made this dissertation possible. Words cannot begin to express the gratitude and love I have for them.

TABLE OF CONTENTS

Chapter

  I. INTRODUCTION 1

 II. REVIEW OF THE LITERATURE 11

       Decision Making Under Risk 13
       Decision Making Under Uncertainty 24
       Portfolio Selection Under Risk and Uncertainty . . . 28
       Risk and Uncertainty in Agriculture 34
       Applications of Risk Theory to Historical
         Issues 39

 III. PRICE, YIELD, AND INCOME VARIABILITY MEASURES 47

       Estimation Procedure 48
         Introduction 48
         The Variate Difference Method 51
         Problems and Limitations 53
         Absolute and Relative Variability
           Measures 55
       Results 57
         General Observations 59
         Random Price Variability of Major Crops 84
         Random Yield Variability of Major Crops 90
         Random Income Variability of Major Crops 96
         Summary 103

  IV. CROP DIVERSIFICATION AND RISK IN THE COTTON SOUTH 105

       Introduction 106
       Measures of Risk for the South 110
         Price Variability of Cotton and Corn 111
         Yield Variability of Cotton and Corn 114
         Gross Income Variability of Cotton
           and Corn 116
       Portfolio Selection 120
         The Model 121
         Estimation Procedure 124
         Results 129

   V. AGRARIAN DISCONTENT AS A RESPONSE TO UNCERTAINTY 139

       Background 140
       Introduction to the Tests 146
       The Cross-Sectional Analysis 147
       The Time-Series Analysis 161
       Results 165

Chapter

VI. SUMMARY AND CONCLUSIONS   173

BIBLIOGRAPHY   183

APPENDIX A: DISCUSSION OF THE DATA AND THE VARIABILITY

MEASURES (Tables A1 through A144 included)   194

APPENDIX B: DISCUSSION OF SUBJECTIVE RISK COEFFICIENT

CALCULATIONS (Tables B1 through B9 included) . . .   298

APPENDIX C: DISCUSSION OF RISK MEASURES AND AGRARIAN

DISCONTENT (Tables C1 through C24 included)   310

LIST OF TABLES

Table		Page
3-1	Selected Crops and Livestock, Representative Sample of States: Ranking by Price and Value Per Head Random Variability Coefficients	60
3-2	Selected Crops and Livestock, Representative Sample of States: Ranking by Yield and Number on Farms Random Variability Coefficients	68
3-3	Selected Crops, Representative Sample of States: Ranking by Gross Income Per Acre Random Variability Coefficients	74
3-4	Five Major Crops (and Hogs), Thirteen Leading Producers: Ranking by Price Random Variability Coefficients	85
3-5	Five Major Crops (and Hogs), Thirteen Leading Producers: Ranking by Yield (and Number on Farms) Random Variability Coefficients	91
3-6	Five Major Crops, Thirteen Leading Producers: Ranking by Gross Income Random Variability Coefficients	98
4-1	Corn and Cotton Price Variability, Southern Cotton Producers: Ranking by Price Random Variability Coefficients of Corn	112
4-2	Corn and Cotton Yield Variability, Southern Cotton Producers: Ranking by Yield Random Variability Coefficients for Corn	115
4-3	Corn and Cotton Gross Income Variability, Southern Cotton Producers: Ranking by Gross Income Random Variability Coefficients for Corn	117
4-4	Estimates of Subjective Risk Coefficients, Southern Cotton Producers: Ranking by States	130
4-5	T-Statistics for the Test of Means of the Subjective Risk Coefficients of All States in a Given Year: Ranking by Year	133
4-6	T-Statistics for the Test of the Mean of the Subjective Risk Coefficients of a Given State Across Time: Ranking by State Name	133

Table   Page

5-1   Wheat, Corn, Oats, and Hay; Fourteen Northern
       and North Central States: Ranking by Price
       Random Variability Coefficients   151

5-2   Wheat, Corn, Oats, and Hay; Fourteen Northern
       and North Central States: Ranking by Yield
       Random Variability Coefficients   153

5-3   Wheat, Corn, Oats, and Hay; Fourteen Northern
       and North Central States: Ranking by Gross
       Income Random Variability Coefficients   156

CHAPTER I

INTRODUCTION

The risks inherent in late nineteenth century agriculture were seldom ignored by contemporary farmers. In fact, farmers have been aware for many decades of the uncertainties and risks in farming. As early as 1872, Charles Stearns, a Southern farmer, wrote a clear description of his knowledge of the uncertainties of crop prices, yields, and income. While Stearns may have been exceptional in his ability to express his knowledge of the risks in farming, he was not alone in possessing this knowledge; it appears that a great number of farmers were aware of these risks.[1]

Scholars of late nineteenth century agriculture also have long been cognizant of the fact that the uncertainties in agriculture influence farmers' behavior. Both Enoch Banks, in 1905, and Rupert Vance, in 1929, included a discussion of risks in their interpretations of postbellum Southern agriculture.[2] And recent textbook accounts of American agricultural development have included the concept of risks as an important factor in the development process. In William Parker's words:

> Western settlement may also be looked at as an economic process, in which risks were borne in return for knowledge--knowledge of climate, soils, and terrain--which might ultimately be put to work in agriculture. The principal obstacle to settlement was not the Indians, the Mexicans, or the federal land policy. It was ignorance: complete uncertainty about the conditions of agriculture in a new area, the risks of Indian attack, insect plagues, climatic variations, and the nature of soils and terrain. Settlement involved an immense learning process by which these uncertainties were converted into defined, experienced, measurable, and even insurable risks . . . . Risks, when known, were not necessarily an obstacle to rapid or skillful settlement. With high risks went high rewards. What was important was that the level of risk should be suited to the gambling instinct of the settlers. The risks had to be brought below the level of Russian roulette . . . (Parker, 1972, p. 376; emphases added).

The importance of the level of uncertainties and risks inherent in nineteenth century agriculture was a central theme in William Parker's explanation of American agricultural development. Soil and climatic variations and market fluctuations led to uncertain crop prices, crop yields, and farm incomes which, in turn, influenced the pattern of agricultural development. On the demand side, Parker argued that the "instability affected different products differently. It was not serious in cotton . . . beef and pork, however, experienced great vicissitudes and the instability of the world wheat market was notorious" (1972, p. 406). From the supply side, he argued that even small fluctuations "in supplies caused by the weather, insects, or disease could set similar short-run fluctuations in motion" (1972, p. 406). According to Parker, the uncertainty of farm prices and yields produced farm incomes which fluctuated sharply. In fact, he suggested that "the instability of farm incomes from year to year" played a major role in the agricultural sector during the late nineteenth century (1972, pp. 407-408). While Parker may have been alone in giving the concept of risks the central role in the interpretation of Western settlement, most other recent textbooks have made readers aware of the existence of instability in agriculture and instability's influence on particular aspects of farmers' behavior.

In discussing changes in American agriculture during the nineteenth century, Vernon Carstensen argued that farmers—who were "an infinitely various group"—commonly faced one particular characteristic of agriculture. In reference to farmers, Carstensen stated, "One certainty they could depend on, if they were willing to face it, was that nothing would remain the same for very long" (1974, p. 12). In other words, farmers faced a rapidly changing environment which, it seems reasonable to assume, led to high levels of uncertainties and risks.

Robert Higgs gave uncertainties of crop yields the central role in his interpretation of farmers' complaints of the last third of the nineteenth century. He described this role as follows:

> Probably greater sources of unrest were the extreme instability
> and consequent unpredictability of farm production. Insects,
> diseases, droughts, prairie fires, floods, hailstorms and
> blizzards—all took their erratic toll from year to year.
> Yields fluctuated madly. Because these random occurrences
> affected differently the various parts of the supplying
> area . . . the farmer could not expect that a low yield
> would necessarily be offset by a high price. It might be,
> but then it might not be . . . but the uncertainty surround-
> ing farm production was substantially greater than that
> associated with most types of nonfarm work. It is reason-
> able to conclude, therefore, that farmers migrated to non-
> farm jobs seeking greater certainty as well as higher real
> incomes (Higgs, 1971, pp. 101-102; emphases added).

Thus, not only did Higgs conclude that a causal relationship existed between uncertainties and risks in agriculture and farmers' behavior, but he also argued that these uncertainties were quite large in magnitude.

Douglass North, in his interpretation of the sources of agrarian unrest, gave the uncertainties inherent in agricultural prices the major causal role in fomenting farmer discontent. He also argued that the magnitude of crop-price fluctuations was extremely great. His views are expressed in the following passage:

> Its causes lay deeper. What was fundamentally at stake in the
> farmer's discontent was, first of all, that he found himself
> competing in a world market in which fluctuations in prices
> created great uncertainty. The bottom might drop out of his
> income because of a bumper crop at the other side of the
> world—in Argentina or Australia. When he suffered from a
> period of drought and poor crops, the higher prices he had
> learned to expect in such a case still might not be forth-
> coming . . . while the international market determined prices
> of wheat, cotton, and some other agricultural commodities,
> many other agricultural foodstuffs and raw materials were
> limited to the U.S. market. The vast domestic market was
> also subject to sharp variations in supply and price (North,
> 1974, pp. 134-136; emphases added).

Again, a scholar was arguing that uncertainty played a causal role in farmers' decision-making processes. He also asserted that these "wide fluctuations in the prices of agricultural commodities" created quite sizeable amounts of uncertainty for farmers (North, 1974, p. 136).

John Peterson and Ralph Gray agreed with most other scholars writing about the conditions of postbellum agriculture, as evidenced by the following passage:

> Farming seldom was easy, and the risks of bad weather were great. Adverse weather took its greatest toll when a poor harvest at home coincided with falling world grain prices brought about by bumper crops on other continents. In addition to the catastrophes of nature that ruined many farmers, the declines in prices from good harvest and over-production brought less income than expected (Peterson and Gray, 1969, p. 291; emphases added).

Again, scholars were suggesting that nature created "great" uncertainties and risks in farming. Furthermore, it was argued that the price declines of the latter part of the nineteenth century were responsible for actual outcomes differing from expected outcomes.

It appeared that the existence of this latter phenomenon was simply another indication of "great" uncertainties in farming. Writing about the economy in general during the late nineteenth century, Milton Friedman and Anna Schwartz supported the existence of this last point. They argued as follows:

> Though declining prices did not prevent a rapid rise in real income over the period as a whole, they gave rise to serious economic and social problems. The price declines affected different groups unevenly and introduced additional elements of uncertainty into the economic scene to which adjustment was necessary (Friedman and Schwartz, 1963, p. 42; emphases added).

To be sure, Friedman and Schwartz were not discussing the issue of over-production in farming (and the consequent price declines), Peterson and

Gray's theme. Instead, they focused their discussion upon the effects
monetary policy had on the economy as a whole. Yet Friedman and Schwartz
implied the existence of a connection between uncertainties (due in part,
to monetary policy) and farmers' behavior when they later stated that "The
decades of the 1880's and the 1890's were notable for political unrest, protest movements, and unsettled conditions" (1963, p. 92). Because they referred to the period of these price declines, this connection becomes clear
when we keep in mind that the "unsettled conditions" to which Friedman and
Schwartz referred predominated in the agricultural sector.

It is interesting to note that a substantive ingredient is lacking
in recent textbook accounts which refer to the level of uncertainties in
agriculture. Knowledge of the actual levels of risk is prerequisite to
any rigorous analysis of the influence of risks on farmers' behavior.
Even so, the necessary empirical work upon which scholars could have
founded their assertions about the magnitude of uncertainties in late
nineteenth century agriculture and its causal role in influencing farmers'
behavior has not been done.

Thus, the primary purpose of this study is to provide objective
measures of the risks associated with various crops and livestock in the
late nineteenth century and to use these estimates, _inter_ _alia_, to study
two important issues in American economic history. Knowledge of these
risks is a necessity to the profession, if analyses of the uncertainties
of postbellum agriculture are to continue. Without this knowledge,
assertions which have little or no empirical content will continue to
be made.

Chapter II is an in-depth review of the risk and uncertainty literature. Theoretical and empirical works are included and a review of the

recent contributions of a small number of econometric historians. The
time spent in discussing the concepts of risk and uncertainty, describing
a number of competing measures of risk, identifying the different approaches
to making decisions under uncertainty, and reviewing the substantive empirical applications of the theory is essential to understanding the direction
and significance of this investigation.

Chapter III contains a discussion of the estimation method used to
derive objective measures of the risk associated with various crops and
livestock and presents the results of the estimation.[3] Estimates of the
uncertainties associated with crop prices, livestock values, crop yields,
livestock numbers, and crop incomes from 1866 to 1909 have been calculated.
General observations on the results of these calculations are included.
These observations consist of a number of comparisons of the general magnitude of the results for several estimated time series and a discussion of
the existence of regional differences in the magnitude of the estimated
risk measures. Finally, the chapter has a detailed discussion of the
computed risks associated with the major crops of several leading producer states.

Because the number of risk estimates computed was quite large--
approximately 1,800 different time series were estimated--a complete list
of the results was not included in the main text of the study. This list
and a detailed analysis of the data (and sources thereof) used in the
computations are in Appendix A. The appendix consists of random variability estimates of the eleven crops--wheat, corn, oats, barley, rye,
buckwheat, potatoes, sweet potatoes, tame hay, tobacco, and cotton--and
the livestock--horses, mules, hogs, sheep, milk cows, and all cattle--

for which annual state data exist. The estimates are listed by state for
each of the forty-eight states.

An issue recently raised by scholars of late nineteenth century
agriculture concerns the influence of risk in shaping the behavior of
postbellum Southern farmers with respect to their choice between cotton
acreage and corn acreage. It has been argued that the apparent increase
in the cotton/corn output ratio between 1860 and 1880 "cannot be explained
by normal changes in economic incentives" but that an explanation can be
found in institutional and historical developments (Wright and Kunreuther,
1975). Advancing a model of how farmers chose their cotton/corn acreage
ratio--which is used to explain shifts in the cotton/corn output ratio,
Wright and Kunreuther argued that institutional changes after the Civil
War "produced a class of 'gambler' farmers" (1975, p. 528). This issue
is investigated in Chapter IV, where a simple risk-management model which
describes farmers' acreage portfolio choice is proposed, and the computed
estimates of the risk associated with cotton and corn crops are used, inter
alia, to test the model. The central concern of Chapter IV is whether the
actual acreage choice of postbellum Southern farmers indicated risk averting or risk preferring behavior.

The estimates of risk associated with a number of crops, which
were computed in this study, also were used to analyze the relationship
between the uncertainties in agriculture and the farmer protest movements
of the last third of the nineteenth century. While the issue of whether
there was an economic basis for farmer discontent is still unresolved, it
has been suggested by a number of economic historians that uncertainties
in agriculture, rather than simply the overall level of economic conditions--

for example, unpredictable income fluctuations rather than overall income levels--were a major source of the agrarian discontent. It is precisely this relationship which is analyzed in Chapter V. The causal relationship between risks and agrarian unrest is analyzed, both for its existence across states and for its existence over time.

In Chapter VI are the summary and conclusions. Also presented are a few suggestions concerning future research possibilities and some general reservations about this study.

FOOTNOTES

[1]For a discussion of Stearns' work and sources of other farmers' remarks concerning risks of nineteenth century agriculture, see McGuire and Higgs (1977, p. 167).

[2]As cited in McGuire and Higgs (1977).

[3]Because the concept of risk and uncertainty in crop production is concerned with unpredictable or "random" variations associated with prices, yields, and income, the empirical procedure employed in this study is one which has as its objective the estimation of random variability of a time series. The variate difference method--the technique employed here--is one such approach. For further discussion, see Chapter III.

CHAPTER II

REVIEW OF THE LITERATURE

The traditional theories of consumer and firm behavior do not include an analysis of uncertain or risky situations. It is generally assumed that outcomes of actions are known with complete certainty. Prices, yields, income, and other variables which affect decisions, usually are assumed to be known with certainty. Yet there is an element of uncertainty about all real-life decision problems. Even though the outcomes of certain actions are predictable, in the sense that probabilities for them can be computed, it appears reasonable to assume that no one can foresee the future perfectly. Therefore, modern decision theory incorporates risk and uncertainty into its models.

Economists following Frank Knight (1921) traditionally distinguish "risk" and "uncertainty" as two different phenomena. When the parameters of the probability distribution of outcomes can be estimated empirically, economists refer to the situation as one of "risk" and when the parameters of the probability distribution of outcomes cannot be established empirically, the situation is referred to as one of "uncertainty." Risk can be defined formally as a state of knowledge in which an action leads to one of a set of outcomes, each outcome occurring with a known probability. Uncertainty is defined as a state of knowledge in which an action leads to a set of outcomes whose probabilities are unknown. Thus, the outcome of an action under certainty is known with a probability of one or zero; the outcome of an action under risk is known with a probability which is less than one but greater than zero; and the probability of an outcome under uncertainty is simply not known.

The plan of this review of the literature is as follows: (1) discussion of decision making under risk, (2) discussion of decision making under uncertainty, (3) review of portfolio selection under risk and uncertainty, (4) applications to agriculture in general, and (5) applications to issues in economic history.

This categorization like other categorizations of topics may appear arbitrary. Some topics in the literature may not fit into one category; whereas, others may fit into more than one category. Nevertheless, this should create no major problems. In addition, because this review of literature is not intended to be a complete coverage of the subject matter, some apparently important topics may be ignored, and, because of overlap, not all of the previously mentioned sections will receive equal treatment.

Decision Making Under Risk

Decision problems involving situations of risk have interested scholars for centuries. The study of decision making under such conditions first appeared in the analysis of a fair gamble. A gamble usually is considered "fair," if the amount (of money) at stake is equal to the expected gain from participating. In this case, the expected gain simply refers to the mathematical operation of taking the average gain. Interested individuals long have asked how much it was worth to participate in a gamble. The answer was that the "fair price" for a gamble was considered to be its expected value. However, Daniel Bernoulli, writing in 1732, argued that for most people the monetary expected value of a gamble was not what they considered to be a "fair price." His introduction of a solution to the now famous St. Petersburg paradox cast considerable doubt on previous analyses of gambling.

The paradox, attributed to Bernoulli's father, is the following:
A coin with the property that the probability of a heads is 1/2 and the
probability of a tails is 1/2 is tossed until a heads appears. If the
head appears on the nth toss, the participant receives 2^n dollars. The
probability of this event--the head occurring on toss n--is 1/2 multiplied
n times. Thus, the participant would have a 50 percent probability of
receiving two dollars, a 25 percent probability of receiving four dollars,
a 12 1/2 percent probability of receiving eight dollars, and so on. The
expected value of the gamble is:

$$2(1/2) + 4(1/4) + 8(1/8) + \ldots = 1 + 1 + 1 + \ldots$$

which equals infinity. Therefore, if the "fair price" for a gamble is its
monetary expected value, an individual would be willing to pay an infinitely
large amount of money to partake in this gamble. This, obviously, is a
rather poor description of behavior. It would appear that people simply do
not behave according to the monetary expected value of such a gamble.

Bernoulli proposed an alternative hypothesis to explain how rational
individuals would make decisions under risk. To save the principle of
behavior according to an expected value, which Bernoulli considered
important, he argued that individuals would look at the expectation of the
intrinsic worth of the monetary value of outcomes rather than averaging the
actual monetary value of the outcomes. It was assumed that the intrinsic
worth of money increases with money at a decreasing rate. Thus, the "fair
price" would be the monetary equivalent of the intrinsic worth (or utility)
expected value, and this sum would have a finite value under Bernoulli's
assumptions.

Even though the justification given by Bernoulli for his hypothesis
is cleverly fashioned, it is an _ad hoc_ hypothesis. It was nearly two

hundred years later before Ramsey (1931) showed that under certain circumstances it is possible to construct a utility index--Bernoulli's
concept of intrinsic worth--for an individual that can be used to predict
his choices in risky situations. Although Ramsey proved the Bernoulli
Principle first, his results were ignored at the time. Most of the
credit for the development of what is now known as the Expected Utility
Hypothesis goes to von-Neumann and Morgenstern (1947). Many scholars
(Marshack, 1950; Herstein and Milnor, 1953; Luce and Raiffa, 1958;
Chipman, 1960) have improved upon the von-Neumann-Morgenstern axiomatic
treatment of utility, but most of expected utility theory, which states
that an individual who conforms to certain axioms will maximize the expected utility of the outcomes of possible actions, still carries their
names.

The axioms of expected utility theory and their implication for
behavior in risky situations can be summarized in the words of Luce and
Raiffa:

> To achieve such a result, it is necessary that the preference
> relation meet certain more or less plausible consistency
> requirements . . . . Among the more important requirements
> were these: preference shall be transitive, i.e., if A is
> preferred to B, and B to C, then A is preferred to C; any
> gamble shall be decomposed into its basic alternatives
> according to the rules of the probability calculus; and if A
> is preferred to B and B to C, then there shall exist a gamble
> involving A and C which is judged indifferent to B. From
> these and other axioms it was shown that one gamble is preferred
> to another if and only if the expected utility of the former is
> larger than the expected utility of the latter. If u is such
> an index, any other is related to it by a linear transformation . . .
> u is called a linear utility function, where "linear" means that
> the utility of a gamble is the expected value of the utilities
> of its components (Luce and Raiffa, 1947, p. 38; emphases in
> original).

Thus, as long as there is a complete and consistent order-preserving ranking of the possible alternatives, there is justification for the central role of expected value.

Given such a preference ranking for an individual, utility numbers are assigned to probability distributions over outcomes such that the ranking is preserved. The individual then chooses from among probability distributions according to the utilities of the distributions. In this way, the tastes (preferences) of the individual, with respect to possible outcomes from his actions and the individual's attitudes toward risk, are captured in a single concept—expected utility.

It is interesting to note that so far no expression concerning the shape of the utility index has been made. The axiomatic treatment of utility implies only that if the values of possible outcomes from an action increase, the utility attached to each possible outcome increases. This is nothing more than an assumption that more is preferred to less. Yet the expected utility attached to a "gamble" also depends on the particular probability distribution of outcomes, where the individual's attitude toward risk represents this individual's preferences with respect to various probability distribtuions. In other words, it is the individual's willingness to accept risks which determines the shape of the utility function over a probability distribution of outcomes, not the theory of expected utility.

Three general shapes of utility functions which represent individuals' attitudes toward risk exist. Suppose we are interested in assigning utility numbers to the monetary outcomes of some gamble; then in the words of James Quirk:

> We say that an individual is risk neutral if, given any gamble
> G, he is indifferent between participating in the gamble G, and
> receiving the expected value of G with certainty. An individual

> is risk averse if he prefers the expected value of G with
> certainty to the gamble G; and he is a risk lover if he prefers
> the gamble G to receiving with certainty the expected value of G.
> These classifications of attitudes toward risk are reflected in
> the measurable utility numbers assigned to payoffs, in that for
> any money payoff strictly between x_1 and x_n the measurable
> utility assigned to a risk averse individual will be higher
> than for a risk neutral individual, which in turn is higher than
> that assigned for a risk lover . . . . Graphically . . . the
> measurable utility curve for a risk averse individual is
> strictly concave, that for a risk neutral individual is linear,
> while the risk lover's utility curve is strictly convex (Quirk,
> 1976, p. 300; emphases in original).

Thus, the classification of the several shapes of utility functions over

monetary outcomes into indicators of attitudes toward risk implies that a

certain measure of risk aversion is the appropriate measure. This point

has been discussed at length in economic literature.

From the time of Bernoulli until recently, it was customary

to view the rate of change of marginal utility of a monetary outcome as a

measure of risk aversion. Thus, if $U(Y)$ is considered a utility function

for money, and because of the assumption that more is preferred to less

$U'(Y)$ is positive, then $U''(Y)$ is used as the measure of risk aversion.

Yet, according to Arrow:

> . . . this suffers from one very severe formal defect. The
> utility function is, after all, a way of representing a
> preference ordering; it is only the latter, and not the former,
> which has behavioral significance. But . . . . the utility
> function is defined only up to positive linear transformations;
> multiplying the utility function by a positive constant or
> adding a constant to it does not change the preference ordering
> represented. Adding a constant to $U(Y)$ does not . . . alter
> $U'(Y)$ and . . . leaves $U''(Y)$ invariant. But multiplying $U(Y)$
> by a positive constant multiplies $U''(Y)$ by the same constant;
> thus, the numerical value of $U''(Y)$ has in itself no
> significance . . . . We thus seek a measure which is based
> on $U''(Y)$ but modified so as to remain invariant under positive
> linear transformations of the utility function (Arrow, 1971, p. 94).

Several authors (Pratt, 1964; Arrow, 1965; Yarri, 1969) showed

that $U''(Y)/U'(Y)$ is such a measure. The ratio $U''(Y)/U'(Y)$ is invariant

to a positive linear transformation while it is still based on the sign
of U''(Y). According to Pratt:

> One feature of [U''(Y)] does have meaning, namely its sign . . . .
> A negative (positive) sign at [Y] implies unwillingness (willingness)
> to accept small, actuarially neutral risks with assets [Y].
> Furthermore, a negative (positive) sign for all [Y] implies
> strict concavity (convexity) and hence unwillingness (willingness)
> to accept any actuarially neutral risk with any assets. The
> absolute magnitude of [U''(Y)] does not in itself have any
> meaning in utility theory, however (Pratt, 1964, p. 127; notation
> in brackets different than original).

Pratt, independent of Arrow's work, also concluded that an appropriate
measure of risk aversion should be some modified version of the rate of
change of marginal utility, namely U''(Y)/U'(Y).[1]

What appear to be, in general, the possible attitudes toward risk
have been discussed, but it has been suggested that a fourth shape might
be a more accurate description of individuals' attitudes toward risk.
It is argued that preference orderings may be more accurately represented
by utility functions which are concave in some interval and convex in
another. Thus, individuals show both risk aversion and risk preference.

This point was first made by Friedman and Savage (1948). They
observed the fact that many individuals who buy insurance also buy lottery
tickets and that obviously, neither activity is "fair" in the sense
described earlier. To be reconciled with the expected utility theory,
these observations imply that the utility function must have an inflection
point near the existing wealth of the individual. The utility function is
concave at lower wealth levels and convex at higher wealth levels.
Friedman and Savage's approach does appear to be a reasonable representation of preference orderings. Indeed, as Borch concluded about the hypothesis, "This is in many ways an attractive assumption. It means that an
improvement over the status quo has a very high utility, and that a very

high loss of utility is assigned to a substantial reduction from the status quo" (1968, p. 37). Borch continued his discussion, however, with an expression of the difficulties inherent in this approach. He argued that if an individual's wealth changes, it is difficult to predict what will happen to the utility function. Will it remain stable or will it shift? For practical consideration Borch asked, "Will a middle-class person who has fire insurance on his home cancel the insurance if he inherits some hundred thousand dollars?" (1968, p. 37).

Considerations such as these lead to another topic in the literature; namely, experiments and observations of economic behavior which have attempted to determine the shape of utility functions representing a typical decision-maker. Only a brief review of the more important contributions to the literature in this area will be presented. It will be accomplished by summarizing the answers provided by Karl Borch to the following questions:

(i) Do the axioms hold in practice, i.e., are they observed by important groups of people who make decisions under uncertainty?

(ii) If the axioms hold, what is the shape of the utility function representing the preference ordering of a typical decision-maker in the different situations which we want to study? (Borch, 1968, p. 62).

For an answer to the first question, Borch referred to experiments carried out by Allais (1953), Mosteller and Nogee (1951), and Davidson, Suppes, and Siegel (1957). The results, in general, showed that when faced with hypothetical risky situations, individuals tended to have inconsistent preferences. It is interesting to note that the experiment carried out by Allais used several well-known economists as subjects. One of them, L. J. Savage, changed his answer when his inconsistencies were pointed out to him! However, referring to the experiment conducted by Allias, Borch suggested:

> It is doubtful if examples of this kind can contribute much to our knowledge about economic behavior under uncertainty. Most people are not used to tossing coins or throwing dice for millions of dollars, and one should probably not attach very much significance to their statements as to how they would make decisions in such situations. One should at least admit that rational people may well make "mistakes" when they state how they would decide in situations which they have never had to consider seriously (Borch, 1968, pp. 65-66).

Nonetheless, Borch, when discussing several other experiments, concluded that they "lead us to doubt the universal validity of the consistency assumptions usually made in economic theory" (1968, p. 72).

The second question which Borch asked must be answered by observing economic behavior in real life situations, because experiments do not give much help in this area. The words of Borch summarize the view held by most economists today:

> They are quite willing to admit that some people like to gamble, so that risk preference undoubtedly exists, but they do not consider this an important element in the economy. The current school of thought is that most respectable people-- the people whose opinions matter--have a risk aversion. The evidence one can quote to support this view is quite over- whelming. Casinos may exist, but they are of no real importance in economic life. The economy is essentially made of "responsible" people who buy insurance and who diversify their investments (Borch, 1968, p. 73).

Thus, even though the specific "shape of the utility function repre- senting the preference ordering of a typical decision-maker" is unknown, in general, it tends to be concave.

If there is a preference ordering over probability distributions of the possible outcomes of an action, it is assumed implicitly that there is a way of measuring risk. In other words, because probability distribu- tions must be ordered according to at least some of their characteristics, the use of any certain characteristics to order outcomes implies that they convey all of the information needed to the decision-maker about the riskiness of each outcome. The crucial question then is, which characteristics of probability distributions should be used to describe riskiness?

The use of a limited number of parameters of probability distributions as a means for ordering outcomes has been widely accepted. Arrow (1971, pp. 24-25) stated that this general idea was popular from the time of Irving Fisher's influence. Fisher (1906) suggested that the desirability of a probability distribution could be determined by knowing the expected value and the standard deviation of possible outcomes. Other scholars accepted the idea that preference ordering of probability distributions may depend solely upon the mean and the standard deviation (or variance), and not the whole distribution (Hicks, 1934; Markowitz, 1952, 1959; Tobin, 1958). Even though he argued that other parameters are to be considered relevant, Marschak (1938) also suggested, at least as a first approximation, the use of the mean and variance for ordering probability distributions.

Where the desirability of probability distributions is based on the mean and variance, the decision-maker will prefer the alternative with a lower variance (if risk aversion is present), as long as alternatives under consideration have identical means; or prefer an alternative with a larger mean, if the variances are identical. As a general rule for ordering risky alternatives, however, this approach has recently been questioned.

Because, in general, expected utility is a function of the whole probability distribution, it appears that preference based on only two parameters would be of limited value. If individuals have a quadratic utility function, then it is true that only the mean and variance are needed to calculate expected utility. Also, if a probability distribution can be described fully by its first two moments (for example, the normal distribution), then this approach will yield valid results.

A number of authors have suggested alternative approaches to the problem of ordering risky alternatives (Hadar and Russell, 1969; Hanoch and Levy, 1969; Rothschild and Stiglitz, 1970). Hadar and Russell (1969) and Hanoch and Levy (1969) suggested independently similar rules for ordering outcomes according to the amount of risk involved. Rothschild and Stiglitz (1970) proposed a condition, which is equivalent to the others, for a special class of distributions (ones with equal means). The major result of these works is that the whole probability density function is needed to rank outcomes in order of riskiness. These authors argued that the identification of risk with any single measure of dispersion is inappropriate, because it does not lead to universally valid results.

How is this result justified? It is based on the stochastic dominance of one distribution over another. The existence of a known probability density function is assumed and, thus, given cumulative probability distributions are known. In addition, expected utility maximization exists. The dominance of one distribution over another exists when the probability of receiving any outcome less than some specified outcome is not larger with the former distribution than with the latter and when the probability of receiving any outcome greater than the specified outcome is not smaller with the former distribution than with the latter (Hadar and Russell, 1969, pp. 32-34).

The above probability conditions will exist, if the value of the cumulative distribution of a preferred outcome never exceeds that of an inferior outcome. Thus, the preferred distributions are said to be stochastically larger than the inferior distributions. This dominance holds under both a strong and a weak condition. The strong condition holds between any two distributions whenever the cumulative distribution of the inferior outcome lies entirely, or partly, above the distribution of the preferred outcome. The weaker condition holds between any two distributions whenever the area under the cumulative distribution of the inferior outcome is equal to, or greater than, the area under the distribution of the preferred outcome (Hadar and Russell, 1969).

It is argued that the ordering of risky outcomes according to the size of cumulative distributions leads to better results, because the whole probability distribution is taken into account. In fact, because the variance of a distribution plays no direct role in either dominance condition, it appears that the variance is an ambiguous measure of risk. More dispersion could be preferred, if it included an upward shift in location of the distribution or more positive asymmetry. In other words, the size of the variance alone might be a poor indication of the riskiness of an outcome (Rothschild and Stiglitz, 1970, pp. 240-242).

Although it is suggested that ordering risky outcomes according to the stochastic dominance approach yields "superior" results because it takes into account the relevance of all moments of a distribution, in practice, the data with which one works may not be amenable to the calculation of much more than simply a mean and variance. Furthermore, the dominance conditions do imply the existence of certain relationships between preference and the first two moments of a distribution. In particular, whether the strong or weak condition is used, the mean of a dominant distribution must be at least as large as that of an inferior distribution (Hadar and Russell, 1969, pp. 27-30). Yet nothing can be said about the relationship between the variance and preference unless the class of distributions is restricted. For the special case of equal means, preference of one distribution over another by a risk averter implies that the preferred distribution must have a smaller variance than the inferior distribution (Hadar and Russell, 1971, pp. 293-294).[2]

In summary, the dominance conditions are a more general approach to the ordering of risky alternatives than those involving comparisons of a limited number of parameters of a distribution. As the moment method restricts either the class of utility functions or the class of distributions, it can yield valid results only in such cases. The practical implications of this discussion may not be apparent yet. However, these issues will be discussed shortly in the section concerned with portfolio selection under risk and uncertainty. The discussion is left until then, because most of the preceding literature grew out of the study of portfolio problems. First, decision making under uncertainty will be discussed.

Decision Making Under Uncertainty

Uncertainty is formally defined as a state of knowledge in which an action leads to a set of outcomes whose probabilities are unknown or have no meaning. In preceding sections, it was assumed that the probabilities of an outcome were known. This meant that individuals would simply study the probability distributions and choose the "best" action. Now that cannot be done. Yet, because we soon will argue that for our purposes the distinction between risk and uncertainty will not be practical, the following discussion of the literature will be rather brief.

The state preference approach to decision making has been developed to describe choice under uncertainty. The essentials of this approach are as follows:

1. There is a set of possible actions, A_m, that an individual can take.

2. There exists a set of mutually exclusive events or "states of nature," E_n, which will take place.

3. A set of possible outcomes or "returns," R_{ij}, exist, which represents the return to an individual for taking action A_i when event E_j occurs.

4. The probabilities of the occurrence of the "states of nature" are unknown to the decision-maker.

It follows from the preceding information that the outcomes of the various actions will have unknown probability distributions which are dependent on the occurrence of the states of nature. The decision problem will be quite different from that under a situation of risk. An individual must choose the "best" action; yet the choice is contingent upon which state of nature occurs.

Several decision rules, which a rational person may follow when making choices under uncertainty, have been suggested in the literature.[3]

1. An individual chooses the action which leads to the largest sum of returns. Thus, the decision-maker will determine

$$\max_{i} \sum_{j=1}^{m} R_{ij}.$$

Any individual who follows this rule assumes, implicitly at least, that all states of nature are equally probable, and chooses the action where the gains totaled over all states are the largest. It is referred to as the Laplace Rule.

2. If an individual is a pessimist who believes that the most
unfavorable event will occur, this person will choose that
action which leads to the largest of all possible minimum
returns. It follows that the decision-maker will determine

$$\max_i \{ \min_j R_{ij} \} \ .$$

This rule would be followed only by individuals who could be characterized
as extreme pessimists. It is referred to as the <u>von-Neumann-Morgenstern</u>
<u>Rule</u>, because it is fundamental to game theory.

3. An individual may decide to weight the largest and the
smallest possible return by some "degree of pessimism" which
he holds. In this case, the decision-maker will determine

$$\max_i \{ \alpha \min_j R_{ij} + (1 - \alpha) \max_j R_{ij} \}$$

where $0 \leq \alpha \leq 1$ represents the individual's "degree of pessimism" or
attitude towards the states of nature. If $\alpha = 1$, the individual is an
extreme pessimist; whereas, if $\alpha = 0$, the decision-maker is an extreme op-
timist. This approach is referred to as the <u>Hurwicz Rule</u>.

4. An individual may worry about the mistake made if the
"wrong" action is chosen. This individual wants to
minimize the regret felt due to a mistake. A regret matrix
with elements

$$\max_j (R_{ij}) - R_{ij} = R*$$

will be computed. The decision-maker then will determine

$$\min_i \{ \max_j R*_{ij} \}$$

where $R*$ represents the difference between each possible return and the largest possible return given each state of nature. This approach, which emphasizes the importance of wrong decisions, is referred to as the Savage Rule.

Although the four decision rules presented may appear reasonable to some people, it is troublesome that the rules could lead to four different actions when applied to the same set of alternatives. In fact, scholars have shown that if it is expected that a "good" decision rule will follow some general conditions which imply a complete preference ordering over the set of actions, then none of the four rules is "good" (Milnor, 1954; Luce and Raiffa, 1957).

Borch (1968) argued that a more promising approach to decision making under uncertainty could be found in the work of L. J. Savage (1954). While the preceding decision rules are independent of the probabilities of the states of nature, Savage's results are not. Savage showed that "beliefs" about the probabilities of the possible states of nature can be incorporated into the analysis. It is argued that once these prior beliefs, which are represented by assigning probabilities to the possible states, are introduced into the decision problem, the gap between situations of uncertainty and situations of risk may have been bridged (Borch, 1968, pp. 202-213). In Borch's words:

> In order to make decisions, a person must establish a preference
> ordering over the sets of actions available to him. Usually it
> will be possible to "describe" this preference ordering in
> several formally different ways. The results of Savage imply
> that the preference ordering is either inconsistent or that it
> is possible to specify a utility function (over the set of
> outcomes) and a probability distribution (over the set of
> events) . . . (Borch, 1968, pp. 86-87).

In other words, if a decision-maker has a consistent preference ordering over the set of actions facing him under uncertainty, it is

possible to specify a unique probability distribution over the states of
nature. But once there is a probability distribution, a situation of
risk exists and the problem can be solved as previously outlined. All
that is needed is a decision rule which represents the individual's attitude toward risk. The characteristics of the probability distribution of
outcomes represent the riskiness of alternative actions.

Thus, we will argue, as does Borch, that there appears to be no
practical reason to distinguish between situations of risk and situations
of uncertainty. Throughout the remainder of this study, it will be
assumed that a decision under risk or uncertainty is simply the act of
choosing an unpredictable or random action which can be represented by
a known probability distribution.

Portfolio Selection Under Risk and Uncertainty

Previous discussion was concerned with an individual's choice of
the "best" probability distribution of outcomes from all possible probability distributions. This section is concerned with an individual's
choice of the "best" mixture of probability distributions from all
available mixtures. We are interested in preference orderings over combinations of risky actions. If there is a preference ordering over all
available alternatives, there will be an ordering over combinations of
these actions, because the mixtures themselves will be risky actions.
In other words, we are interested in how individuals select their
"optimal" portfolio under uncertainty (Borch, 1968, p. 47).

Because a large part of the literature previously discussed was a
development of portfolio selection problems and because the selection of
an optimal portfolio under uncertainty can be analyzed in a manner similar

to the choice of a single risky alternative, the preceding topics are applicable here. The main interest in this section is to review the literature concerning the major controversy surrounding portfolio selection, namely that involving the use of the mean-variance approach to portfolio choice under risk.

An excellent summary of the foundations of the mean-variance approach was provided by Josef Hadar and William Russell:

> From its very inception, the theory of portfolio choice was based on the principle of minimizing the variance of any portfolio consisting of a number of assets with equal means. This principle was derived from the proposition that, between any two distributions with equal means, a risk averter always prefers the distribution with the lower variance. It is this approach which forms the basis of the works on portfolio choice . . . (Hadar and Russell, 1971, p. 288).

This approach was first suggested by Harry Markowitz (1952, 1959) and later followed by others.[4]

Prior to Markowitz, decision-theory under uncertainty dealt with the problem of the selection of the appropriate single risky asset among many investment opportunities. His suggestion, in 1952, was that investment decisions should be studied in the context of determining the appropriate portfolio of assets among many available portfolios. Thus, suppose we have n different risky assets; we can then outline the model suggested by Markowitz. Let

E_i = mean (or expected) return on asset i;

V_i = variance of return on asset i;

Cov_{ij} = covariance of returns on assets i and j;

q_i = proportion of portfolio held in asset i.

Then, any portfolio consisting of n assets will have expected return

$$E = \sum_{i=1}^{n} q_i E_i$$

and variance

$$V = \sum_{i=1}^{n} q_i^2 V_i + 2 \sum_{i \neq j}^{n} Cov_{ij} q_i q_j.$$

To solve the problem, Markowitz assumed that a utility function U(E,V) existed such that

$$\frac{dU}{dE} > 0 \text{ and } \frac{dU}{dV} < 0 \text{ are satisfied.}$$

An individual then starts with any arbitrarily acceptable expected return, E*, and determines which portfolio of all portfolios with E* return has the least variance. The process is repeated for all conceivable values of E* so as to have a set of portfolios—each of which has minimum variance for a given expected return. The solution is to determine the unique values of q_1 to q_n (for each E*) which would represent, in the words of Markowitz, an "efficient portfolio" with expected return E*. This set of efficient portfolios is referred to as a mean-variance (E-V) frontier.

Markowitz suggested that once the set of efficient portfolios—those portfolios containing minimum variance for each expected return—is known, the decision-maker can study it and determine for himself which portfolio satisfies his "preferences with respect to risk and return" (1959, p. 23). In other words, the final decision, according to Markowitz, is purely a subjective one. The manner in which risk is weighted is not specified uniquely in the model.

The mean-variance approach to decision making under uncertainty became widely popular among scholars after its introduction by Markowitz in 1952. Freund (1956) used a variant of the Markowitz model to determine an optimum combination of crops—a portfolio—for a typical Eastern North Carolina farm. He argued that by introducing risk into a programming model for a typical farm, one is better able to make recommendations as to the "best" acreage portfolio. Tobin (1958) used the approach as a foundation for a theory of the demand for money. He argued that a theory of risk averting behavior, based on a mean-variance analysis, could be used to explain the existence of an inverse relationship between the rate of interest and the demand for money. Farrar (1962) developed a decision model, based on the mean and variance of returns, to study the attitudes toward risk of twenty-three mutual funds. He suggested that the Markowitz model could be used to show that the stated intentions of mutual funds with respect to risk and the portfolios actually held by them are, in general, consistent. Cheung (1969) used a mean-variance model and the presence of risk aversion, among other things, to explain the choice of contractual arrangements. In particular, he argued that share tenancy in agriculture can be explained, in part, by risk averting behavior--where agricultural risk is measured by income variance. Others also have used the Markowitz model or variants of it to study portfolio selection under risk (Sharpe, 1963, 1964; Baumol, 1963; Fama, 1965).

Though a considerable body of literature in the area of portfolio selection has taken the mean-variance approach, the status of the approach has not been without controversy. A number of authors have shown that the use of the first two moments of a probability distribution for ordering uncertain outcomes is valid only if the utility functions or distributions are restricted (Borch, 1969; Feldstein, 1969).

These criticisms of the mean-variance approach were a development of expected utility theory. Unlike Markowitz's, most of the work that followed assumed expected utility maximizing behavior. Yet, if this assumption is held, then the use of the mean and variance as the only two

parameters that matter implies some severe restrictions. The probability
distributions under study could be restricted only to those which belong
to a two-parameter family (Borch, 1969). In particular, any two-parameter
family where only the mean and variance was needed to describe the whole
probability distribution could be used--for example, the normal.

If the preceding restriction does not hold and mean-variance
approach with expected utility theory is used, the utility function under
study must be restricted to a quadratic. However, a quadratic utility
function must be decreasing in some interval. Thus, higher levels of
wealth will decrease utility. The only possibility for using a quadratic
is to place further restrictions on the function: it must be bounded to
its increasing range only (Feldstein, 1969).

Restrictions on both the class of utility functions and the class
of probability distributions have come under heavy criticism. It is
implied that not only must a quadratic function be restricted to its
rising portion, but absolute risk aversion also must be increasing--as
wealth increases individuals become more risk averse. For these reasons,
Arrow (1971, pp. 90-109) considered the quadratic function completely
unaccepteable. Others have argued that the normality assumption is
equally suspect, because it too narrowly restricts the types of uncertain
prospects one can study (Rothschild and Stiglitz, 1970, p. 241).

A large number of studies (Hadar and Russell, 1971, 1974;
Rothschild and Stiglitz, 1971; and others) on portfolio theory developed
out of the stochastic dominance literature as an alternative approach to
mean-variance analysis. Their major contribution was the presentation of
rules for ordering uncertain portfolios when expected utility maximizing
behavior was assumed. Because the rules do not depend on restrictive

assumptions, it is argued that they generally are more valid than other orderings. In order for preference to be determined, only the cumulative probability distribution of each portfolio under study is needed. As mentioned earlier, these rules are presented under both a strong condition—which applies essentially to all utility functions—and a weak condition—which applies to those functions exhibiting risk aversion—where both dominance conditions imply certain relationships between preference and the moments of a probability distribution.

It is interesting to note that, although the various criticisms do apply to the works of Tobin (1958), Farrar (1962), and others, they do not apply strictly to the work of Markowitz (Baumol, 1963). Tobin, Farrar, and others used expected utility theory as a basis for their work; whereas, Markowitz did not. Markowitz, because he believed expected utility theory to be controversial, simply assumed that a utility function exists where the mean and variance of returns of a portfolio are direct arguments in the function (Borch, 1968, p. 50). Thus, the model rests on the assumption that an individual will look at only two parameters of an uncertain portfolio.

Whether or not the theoretical distribution of uncertain outcomes are described by only their mean and variance was irrelevant to Markowitz. He felt that individuals simply may not look beyond the mean and the variance when ordering risky portfolios, and some agreement with this line of reasoning exists in the literature.

Though often critical of E-V analysis, both Borch (1968) and Feldstein (1969) believed that it may be a reasonable first approximation of behavior under uncertainty. It may be that making a "statistical mistake" leads individuals to believe that the mean and variance can be used to order uncertain portfolios (Feldstein, 1969, p. 10), or that "people find it easier to compute than to state a complete and consistent set of preferences" (Borch, 1968, p. 52). Also, it may be that the choices facing an individual really do not have much "skewness" and, thus, the mean and variance of returns are a reasonably accurate description of the uncertain portfolios (Borch, 1968, pp. 60-61). As justification for using the E-V approach it also may be claimed that, as far as individual decision-makers are concerned, information on higher order moments of the probability distribution of returns is prohibitively costly to obtain (Gould, 1974).[5] Thus, the mean and variance of returns (for these individuals) do the "best" job of describing uncertain portfolios.

A mean-variance approach was the method of analysis employed to study the question concerning the cotton/corn acreage portfolios of post-bellum Southern farmers discussed in Chapter IV. A mean-variance model of portfolio choice may not lead to universally valid results, but Chapter IV will show that for our purposes it is a reasonable first approximation of behavior under uncertainty.

Risk and Uncertainty In Agriculture

The body of literature in the area of agricultural economics concerning risk and uncertainty is considerable. Because the literature is so extensive, little more than a brief summary of the general direction the literature has taken will be presented. There will be, however, a more detailed review of the work in agricultural economics upon which this study has drawn heavily, namely the work of Harold Carter and Gerald Dean (1960).

During the 1940's, agricultural economists such as O. H. Brownlee, Earl Heady, D. Gale Johnson, and T. W. Schultz began exploring the problems of risk and uncertainty in farming.[6] The efforts of these scholars were directed more at identifying the issues than anything else, but their studies were the forerunners of more concrete analyses. Writing in the 1940's, these scholars called for the beginning of economic analyses of decision making in the face of risks and uncertainties. It was argued in particular that applied work, to the area of agriculture, was needed.

Kling (1942) was one of the first to do an empirical study of risks in agriculture. His work presented estimates of the risks involved in growing truck crops in the United States from 1918 to 1940. The estimates were computed by calculating the standard deviation of the first differences of annual observations of prices, yields, and income per acre for fifteen crops. It was argued that first differences should be used so as to minimize secular trend. Kling suggested that farmers could use this information about the "degree of annual variability" associated with various crops in their decisions concerning choice of crops (1942, p. 698).

Heady, in 1952, made the first attempt to detail the relationship between diversification principles and income variability. He presented a framework for combining crops, as a means of lessening income variability, by taking into account the relationship between variances and covariances of income associated with different crops. Included in the analysis were some empirical examples of how the diversification principles work (Heady, 1952, pp. 488-490). Since Heady's seminal analysis, a number of scholars have provided empirical estimates of the amount of variability--used as a proxy for riskiness--in prices, yields, and income per acre associated with various crop enterprises (Jensen and Sundquist, 1955; Swanson, 1957; Carter and Dean, 1960; Grossman and Headley, 1965). Because these studies

accepted the appropriateness of measuring risk in terms of a single
parameter of a probability distribution, their efforts were directed at
finding an appropriate empirical measure of dispersion. An estimate of
the variance of the various series under study usually was chosen.

The study by Carter and Dean (1960), a thorough empirical study,
is used as the framework for Chapter III. The objectives of Carter and
Dean were the estimation of "the degree of variability in yields, prices,
and incomes associated with various types of crop production in California"
and the investigation of "the relationships between stability and level
of farm income from particular cropping systems" (Carter and Dean, 1960,
p. 176). Their computed estimates of variability might not be identical
to the concepts of risk and uncertainty, but they argued that the estimates
are objective measures which approximate the degree of risk in prices,
yields, and income associated with different crops.

While the Carter and Dean approach was essentially a mean-variance
analysis of the risks and uncertainties in California agriculture from
1918 to 1957, it did present an interesting way of looking at the problem.
The argument was that farmers are interested in the unpredictable or
"random" variability associated with various crops as an approximation
to the degree of risks. Because it would be unreasonable to assume that
farmers had no knowledge about long-run economic or physical trends in
agriculture, only part of the total variation in prices, yields, or income
could be viewed as random to an individual farmer. Thus, Carter and Dean
(1960, p. 177) argued that only deviations from current levels of prices,
yields, or income should be used as a measure of risk.

The procedure they suggested for determining the current level of
a time series (and, thus, estimating the variability of the random element)

was Tintner's variate difference method. The authors perferred that method over alternative procedures, because it does not require advance specification of rigid functions. They argued that advance specification implicitly assumes knowledge of economic variables which actually may not be available.[7]

The empirical estimates of variability in the Carter and Dean study were expressed in both absolute and relative measures, the former being an estimate of the random variance of a time series and the latter being the former expressed as a percentage of the mean. These estimates were used to test several diversification principles as ways of lessening risks in farming. An assumption implicit throughout their analysis is that future expectations are based on past experience (Carter and Dean, 1960, p. 175).

Other studies, investigating a variety of topics concerning agriculture, employed techniques similar to those used by Carter and Dean to determine the degree of risk associated with different crop enterprises (Grossman and Headley, 1965; Moore, 1965; Moore and Snyder, 1969). Grossman and Headley presented variability estimates associated with selected crops for Illinois during the first half of the twentieth century. Those estimates were used as a basis for an E-V analysis of acreage decisions made by Illinois farmers. Moore, in 1965, computed current income variability associated with eight crops for farmers in Fresno County, California. He used the estimates to study the relationship between farm size and income variability. His analysis showed that an inverse relationship may exist between increased risks and both the rate of expansion of farm size and capital accumulation. Moore and Snyder, writing in 1969, applied Tintner's variate difference method to several income series of vegetable crops grown in the Salinas Valley of California. The estimates were used in an E-V framework in which the objective was the maximization of long term income. Moore and Snyder concluded that some acreage portfolios not included in standard mean-variance models, where short term maximization is assumed, should be considered as alternatives when maximization of long term income is the case.

In addition to articles estimating the amount of variability associated with various crops and crop enterprises, much has been written about use of the Markowitz model and variants of this model for solving farm management problems. Methods have been developed to simplify farm planning by reducing the set of efficient portfolios facing the farmer. This can be accomplished by comparing confidence limits of the expected income from the possible portfolio combinations (Baumol, 1963). Also, quadratic programming can be utilized to determine an E-V (mean-variance) frontier. This frontier can, in turn, be used for the selection of a crop enterprise (Stovall, 1966).

Other authors applied the quadratic programming model and presented suggestions for determining a farm plan under risk (Scott and Baker, 1972). A number of similar models, using linear programming techniques, were developed as alternatives to quadratic programming, and had as their objectives the reduction of computational costs (Hazell, 1971; Thomson and Hazell, 1972; Thomas, Blakeslee, Rodgers, and Whittlesey, 1972). Still other approaches employed linear models as alternatives to the standard mean-variance analysis. One such model viewed farmers as having multiple goals and discontinuous preferences (Boussard and Petit, 1967); whereas, others were interested simply in introducing risks and uncertainties into standard input-output models (Madansky, 1962; Maruyama, 1972).

Several agricultural economists had an altogether different approach. In the belief that risk and uncertainty are two different phenomena, some scholars presented models which utilize the "states of nature" approach to decision-making under uncertainty to solve farm management problems (Dean, Finch, and Petit, 1966; Eidman, Dean, and Carter, 1967; Cocks, 1968; Bullock and Logan, 1969). Probably the most complete treatment of this approach, one that also shows how partial information on the probabilities of the states of nature can be utilized, was a paper presented by Rae (1971).

Even though the foregoing models, which utilize the state preference approach, are quite proper, they do not appear useful for purposes of this study, which takes the approach that the probabilities of outcomes can be established empirically. In conclusion, it should be mentioned that applications of the stochastic dominance approach to farm management problems were not to be found in the literature.

Applications of Risk Theory to Historical Issues

Recently, a number of economic historians incorporated risk into their explanations of several historical issues (Reid, 1973; Higgs, 1973, 1974; Wright and Kunreuther, 1975). All were concerned with issues pertaining to postbellum Southern agriculture--the concern of Chapter IV. In fact, Chapter IV of this study presents an alternative approach to the acreage management problem contained in Wright and Kunreuther (1975). While the explicit recognition of the importance of risk in shaping individuals' behavior is long overdue among historians, much of the recent explanations contains faulty theoretical and statistical procedures.

The paper presented by Reid (1973) investigated the historical issue concerning postbellum Southern economic stagnation. He is interested, in particular, in whether the postbellum stagnation could be attributed, among other things, to the rise of sharecropping as many have claimed. This issue was analyzed by proposing an economic basis for sharecropping. Reid argued that "the rise of tenancy should have _increased_ southern agricultural productivity, _cet. par._ Thus the postbellum fall in southern agricultural productivity cannot be directly attributed to the rise of tenancy" (1973, p. 107; emphases in original).

The introduction of risk into Reid's explanation of sharecropping came in a unique manner. Unlike others (Cheung, 1969; Higgs, 1973), Reid argued that risk dispersion is _not_ the main economic function of sharecropping, and that agricultural uncertainty cannot be used to explain the geographical dispersion of share tenancy. He suggested that sharecropping is used to "facilitate aggregate risk reduction. By encouraging both parties to the tenure arrangement to similarly respond to revisions in expectation over the crop season . . . sharecropping minimizes expected total risk . . . " (1973, p. 126). It is this part of Reid's argument in which we are interested. Even though his central proposition concerned the economic advantages of sharecropping, that is not the concern here.

Looking at the corn and cotton crop for Southern states, Reid based his argument on evidence "that the ranking by proportion of each crop sharecropped was the _opposite_ of the relative uncertainty ranking in half of the considered cases" (1973, p. 121; emphasis in original). Yet his comparison of the relative riskiness of cotton and corn was based solely on yield variance. It is the uncertainty of income which concerns a farmer—whether a landlord or tenant. However, not only did Reid ignore price variability, but he also ignored the possibility of covariability between prices and yields. In addition, cotton and corn were treated separately, and that should not be the case. Because both crops are grown on the same farm, it is actually the riskiness of the whole portfolio of crops which concerns the farmer. Thus, one needs a measure of the variability of the whole package which takes into account the income variance of both crops and the income covariability between the two crops.

We are not suggesting that there is necessarily a connection between risk dispersion and share tenancy. Our only argument is that the necessary empirical work to test for such a relationship is not contained in Reid's (1973) paper. As far as "aggregate risk reduction" is concerned, Reid presented no empirical support for that conjecture.

Higgs (1973, 1974) offered a different explanation for an economic basis of sharecropping than Reid. Following Cheung (1969), Higgs argued that the main impetus for sharecropping _is_ risk dispersion. He believed that regional patterns in agricultural uncertainty in the postbellum South may help to explain the geographical patterns of share tenancy. However, the empirical support presented by Higgs (1973, p. 158) suffered from many of the same defects as Reid's work.

The assertion presented is that "the yield risk and the percentage of rented acreage under share contracts are directly related" (Higgs, 1973, p. 157). Yet, if the economic function of sharecropping is to facilitate risk dispersion between landlords and tenants, it appears reasonable that one should be interested in the income uncertainty associated with various crops and crop enterprises—not simply the yield uncertainty. To be fair,

Higgs (1973, p. 154) argued that contractual arrangements are a function of "natural" risk, and that the yield variance is a reasonable proxy for the uncertainty of nature. Thus, the question is, is sharecropping used solely for dispersing natural risk or is it also used to disperse the risks of the market place? We believe that the assumption of risk aversion should apply to both types of risk. In addition, the estimated regression equations offered by Higgs (1973, p. 157) were computed for corn and cotton separately. They should have been treated together as the whole portfolio of crops, as mentioned earlier.

The analysis of acreage management in postbellum Southern agriculture contained in Wright and Kunreuther (1975) is a sophisticated application of risk theory to an historical issue. Their main concern was to investigate the economic basis for the apparent rise in the cotton/corn output ratio between 1860 and 1880. Wright and Kunreuther did not believe that standard economic incentives could be used to explain this change, because there were no normal changes in these incentives between the two dates. This argument was based on the fact that neither relative prices nor crop yields changed much between 1860 and 1880. The explanation for the rise in the cotton/corn output ratio advanced by Wright and Kunreuther was that "institutional and historical developments" between 1860 and 1880 were such that farmers were forced to "gamble" on cotton (1975, p. 528).

The model proposed by Wright and Kunreuther to explain the apparent shift into cotton production during the postbellum period was a variant of "safety-first" models used by developmental economists. The specific model and its implications can best be described in their words:

The safety-first hypothesis is that the farmer maximizes
expected earnings after assuring that the risk of falling
below Z* (the critical yield level per acre which will just
produce minimum tolerable consumption) is less than α*
(the risk level which the farmer is willing to tolerate).
But suppose there is <u>no</u> allocation of acreage which satisfies
this constraint. In that case, the farmer must . . . concen-
trate all his efforts on maximizing his <u>chances</u> of achieving
Z* . . . . (H)e is now "forced" to gamble on the cash crop
in order to have any chance of achieving his critical target
yield . . . . (W)e do not argue that small farms of the
South were literally too small to feed a typical family.
But we do maintain that a variety of changes combined to
raise Z* and α*, creating a class of farmers for whom aspirations
and circumstances dictated a quasi-voluntary decision to "gamble"
by planting a large share of their limited acreage in cotton
(Wright and Kunreuther, 1975, pp. 539-540, 531; emphases in
original).

Using corn output as a proxy for foodstuffs and cotton as the cash crop,

the authors explained the rise in the cotton/corn output ratio as a response

to "gambling" behavior, which was created by institutional developments

between 1860 and 1880.

The significance of the rise in the cotton/corn output ratio is

questionable; however, it is still important to take a critical look at the

Wright and Kunreuther approach.[8] They presented an <u>acreage</u> management

model—where risk is incorporated as an explanatory variable—as a basis

for their explanation of the change in the cotton/corn <u>output</u> ratio.

While some of the properties of the model might be suspect, the central

concern here is with the empirical evidence presented to support their

approach.[9]

Wright and Kunreuther's argument was that the cotton/corn mix in

1880 implied gambling behavior on the part of farmers. Yet their basis for

this belief was data on the mean return and the standard deviation of returns

computed for corn and cotton separately from annual data for the period

1866-1900. Because the standard deviation associated with cotton is greater than that associated with corn, and likewise for mean returns, the authors concluded that cotton is a riskier crop. It was argued, therefore, that because the cotton/corn output ratio appeared to be higher in 1880 than in 1860, farmers at the later date must have been gambling on cotton. However, that was not shown.

Wright and Kunreuther not only ignored the relative magnitudes of the standard deviations, they argued that a particular portfolio of crops was risky without evidence on the variability of that portfolio. By calculating variability measures for the two crops separately, they ignored covariability between cotton and corn returns. For this reason, one cannot say that a portfolio containing more cotton is riskier than another portfolio with less cotton, simply because cotton may be a riskier crop than corn. In addition, it is not proper to use a variability measure computed for the period 1866-1900 to describe the riskiness of crops grown in 1880 unless there was homogeneity of variance before and after 1880, and the necessary calculations were not done. Lastly, the concept of risk usually refers to the unpredictable or "random" variability associated with a series. By ignoring systematic movements in the data, Wright and Kunreuther implicitly assumed that farmers view all fluctuations from the long-run mean as unpredictable. In other words, they assumed that farmers are completely ignorant about existing trends. That is an unreasonable assumption.

Although the research efforts of Reid, Higgs, and Wright and Kunreuther should not be underestimated, it appears that the objective measures of risk presented in this study will prevent similar measurement problems for economic historians who are interested in late nineteenth century agriculture. They will, at least, provide a useful set of data on agricultural uncertainty for other historians.

FOOTNOTES

[1]It is interesting to note that both Arrow (1965) and Pratt (1964) reached their conclusions with an assumption of expected utility maximizing behavior while Yaari (1969) did not start with such an assumption. Nevertheless, the risk aversion measure proposed by Yaari (1969, p. 318) was simply "proportional to the Arrow-Pratt measure" and led to similar conclusions about behavior.

[2]This conclusion about the case of distributions with equal means is contrary to that argued by Rothschild and Stiglitz (1970). They claimed that "if X and Y have the same mean, X may have a lower variance and yet Y will be preferred to X by some risk averse individuals" (1970, p. 241). Nevertheless, Hadar and Russell (1971) showed that Rothschild and Stiglitz were incorrect for any nonnegative random variable.

[3]The discussion of decision rules under uncertainty is a condensed version of that presented in Chapter 7 of Borch (1968).

[4]Even though Fisher, Hicks, Marschak, and others suggested the use of the variance as a risk measure much earlier, Markowitz was the first to suggest its use in portfolio problems. Freund (1956), Tobin (1958), and Farrar (1962) patterned their work on the seminal analysis of Markowitz.

[5]Gould did not claim that information costs may justify the use of an E-V approach. He merely discussed the possibility that these costs may alter decision-makers' behavior.

[6]For an excellent survey of the early literature on risk and uncertainty applied to farm management problems, see Johnson (1952). The noted works are as cited there.

[7]A detailed discussion of Tintner's variate difference method is presented in Chapter III of this study.

[8]McGuire and Higgs (1977, pp. 168-170) have identified erroneous evidence presented in Wright and Kunreuther (1975) which, when corrected, implies that there may have been little or no rise in the cotton/corn output ratio between 1860 and 1880.

[9]The Wright and Kunreuther model violates the continuity assumption usually made in economic theory. In addition, it may be inapplicable to the postbellum South on empirical grounds. For a detailed consideration of the model, see McGuire and Higgs (1977, pp. 170-172).

CHAPTER III

PRICE, YIELD, AND INCOME VARIABILITY MEASURES

Estimation Procedure

Introduction

Risk and uncertainty were distinguished as two different phenomena by Knight (1921). Risk refers to situations in which parameters of the probability distribution of outcomes can be estimated so as to be insurable. Uncertainty refers to situations where parameters of the probability distribution of outcomes cannot be measured empirically. In other words, "risk" can be measured quantitatively by looking at the variability of outcomes; whereas, "uncertainty" cannot.

This study assumes that certain parameters (mean and variance) of the distributions of prices, yields, and incomes can be established empirically. It also is assumed that these two moments accurately describe the probability distribution of outcomes of prices, yields, and incomes. The variance of the "random" portion of each time series is used as the measure of the variability associated with various crops, cropping systems, and livestock. The traditional distinction between "risk" and "uncertainty" (as being different phenomena) will not be observed here.

Although empirical variability estimates are not necessarily identical with the traditional concept of risk or uncertainty, they are objective measures of past variability. They can be interpreted as a first approximation to the amount of risk which a farmer might expect to face. An implicit assumption is, thus, that the farmer bases his future expectations on his past experience.

Three types of crop variability and two types of livestock variability are considered in this study:

Price Variability. Crop prices are subject to variability from numerous sources. Any factor that causes market demand and/or supply conditions to fluctuate will affect price stability.

Yield Variability. Yield variability arises from uncertain weather conditions, disease, insect problems, resource availability, and technological change.

Income Variability. Income variability per acre arises from the interaction of crop yield per acre and crop prices relative to costs. Thus, variability in income is of primary interest to all farmers.

Value Per Head Variability. Livestock values are subject to variability from any factor that causes market demand and/or supply conditions to fluctuate.

Number on Farms Variability. Variability associated with livestock arises from uncertain weather conditions (for example, droughts), disease, plague problems, and other exogenous factors.

Variability in farming stems from the fact that yields, prices, and income are influenced by numerous factors in an unpredictable or "random" manner. An estimate of the portion of the total variation in price, yield, and income that is unpredictable or "random" from the standpoint of the individual farmer is desired. A variability estimate which uses the variance of an original time series implies that farmers view any deviation from the long-run mean as unpredictable or random. Such a situation implies that farmers have "no knowledge" about certain long-run trends--such as price levels, business cycles, and technological change. It seems more reasonable to assume that farmers have some knowledge about long-run physical and economic trends. Therefore, a more realistic view of the random element would be a variability estimate which uses deviations from the current mean, instead of deviations from the long-run mean. It is only this "random" portion of total variability that is relevant to farmers' response to risk. If a consequence is predictable (known), then it is not "risky" or "uncertain."

There are several competing empirical procedures for determining the actual "current level" of any time series. Thus, there are a number of methods for computing the deviations from the current mean (Yule and Kendall, 1965, pp. 610-637).

One method of finding deviations from the current mean is to estimate the current level of a time series by the use of moving averages. Deviations from this level are then used as a measure of variability. A second procedure is to approximate the current level by simple curve fitting, in which a fitted trend line is the estimate of the current mean. Thus, deviations from the trend line are considered as the random element of the time series. A third method assumes that the current level is the same as the previous year, and, thus, first differences of the data are used as estimates of the unpredictable component. Lastly, a time series can be adjusted by some general index so as to obtain "real" values of the series. The researcher then uses variability estimates derived from deviations from the long-run mean of the adjusted series.

These alternative procedures for calculating the current level of a time series have advantages and disadvantages. The one common disadvantage is that the methods require a priori specification of some type of rigid function. A trend removal method which allows flexibility is desired, because a priori information about functional forms is not always known. Because our interest is in the random component rather than the trend, a method which has as its objective the determination of random deviations and the estimate of their variance also is desired. The variate difference method is one such procedure which meets these objectives and allows for flexibility (Tintner, 1940; Yule and Kendall, 1965, pp. 631-633). Therefore, the variate difference method was chosen for this study.

The Variate Difference Method

The fundamental assumption of this technique is that the systematic and random components of a time series exist and are additive. The systematic component of a time series is ordered in time, such that consecutive observations are positively autocorrelated. This component can be referred to as the mathematical expectation. The random element of a time series consists of observations which are neither autocorrelated, nor correlated with observations of the mathematical expectation. The variate difference method separates the systematic component from the random component by successive finite differencing. The method avoids unnecessary assumptions about the specific functional form of the mathematical expectation. All that need be assumed is that the systematic element of a time series can be described by an unspecified polynomial or other type of mathematical function of the variable time. Tintner (1940) showed that the Nth finite difference of a polynomial of degree N is constant and the N+1, N+2, and higher-order finite differences vanish. However, this does not hold for the random component, which cannot be reduced by finite differencing, because it is not ordered in time. Tintner stated the fundamental idea as follows:

> We begin with the assumption that the economic time series in question consists of two parts: a "smooth" part which we shall call the mathematical expectation and which we assume is the result of the more permanent economic and social factors, and a random part . . . . It may be considered the result of the nonpermanent or less permanent factors in economic life . . . . We assume that the mathematical expectation and the random element are connected by addition. The variance of the entire

series can hence be split up into one part which comes from
the mathematical expectation and another part which is the
random variance. It has already been pointed out that the
smooth part or the mathematical expectation can be eliminated,
or at least reduced to any desired degree, by successive
finite differencing. Hence the part of the variance that results
from the mathematical expectation can be reduced . . . . The
variance of the finite difference of any order is again equal
to the sum of the variance of the difference of the random
element. The first component is reduced . . . by successive
finite differencing, the second component is not changed . . .
(1940, p. 32).

From this position, Tintner explained his procedure for trend removal as

the following:

We ask ourselves the following question: Beginning from which
finite difference k_o can we assume that the mathematical
expectation has been eliminated to a considerable degree and
that we are left approximately with only the random element?
The solution of this problem results from the following
consideration: If we have a series which consists only of
the random element, then the variances of the successive series
of finite differences are equal, . . . This is self-evident
from the fact that the series . . . is not ordered in time.
Hence the variance of its first and higher differences must be
the same as the variance of the original random series . . . . If
we find a certain finite difference of the order k_o such that
the variance of the k_o^{th} difference is equal to the variance
of the $(k_o+1)^{th}$ difference . . . , etc., then . . . we have
eliminated the mathematical expectation . . . by taking k_o
differences . . . . Hence, we will show that it is required
only that the differences between the variances of two successive
series of finite differences be smaller than three times its
standard error. This is sufficient from the point of view of
the theory of probability in assuring us reasonably well
that . . . only traces of the mathematical expectation are
left . . . . We can base this argument on Tchebycheff's
inequality (1940, pp. 32-33).

Thus, the procedure consists of calculating the variance of the original

series and for each of the series of successive finite differences.

Comparisons of variances are made until the criterion is met, then the

finite differencing has been carried far enough to yield a valid variance

estimate of the random element.

Problems and Limitations

The portion of the total variance of a time series which results from the mathematical expectation can be reduced and/or eliminated by successive finite differencing only for a "smooth" series. That is, one must be able to approximate the systematic component of a series (where consecutive items are positively correlated) by polynomials of the variable time. As long as this approximation holds true locally, finite differencing can be used to separate the variance of the mathematical expectation from the variance of the random element. However, if consecutive items in a time series show strong negative correlation, then the variate difference method is not applicable (Tintner, 1940). This is the case of the extreme "zigzag" series. Series of this type have been omitted here as they cannot be detrended with the technique employed.

The variate difference method is not applicable to any time series which consists of periodic oscillations of short duration, because such periodic fluctuations are not smooth, but are of the extreme zigzag type. However, fluctuations of this nature do not occur frequently in economic time series. Seasonal variation of a series would not fit into the preceding category. Fluctuations which are seasonal in nature generally do not affect the results of the variate difference method, however (Tintner, 1940).

The criterion for the stability of the variances of the series of finite differences is strictly justified only if the series is long enough, because the criterion is based on the theory of large samples and does not hold for small samples. For a short time series, there is no clear-cut rule to determine whether the variance from the systematic component of the series has been eliminated so as to be left with an estimate of the variance of the random element only. Because the choice of which difference provides the best estimate of the random component is not clear on statistical grounds, the researcher must use his own judgment for a short times series. This is not a major problem here. However, it is noted that, where applicable, a different choice criterion is used.

Another problem unique to the variate difference method is that repeated finite differencing of an already random series introduces artificial autocorrelations into the series and, hence, prevents one from using this method to determine the "random" element of the series. The variate difference method is applicable only if there is no autocorrelation in the random component of a time series (Tintner, 1940). However, this need not be a problem. If a random series exists in which every item is entirely independent of every other, it can be detected by estimating serial correlation coefficients for each series and analyzing the results. If it is determined that the items in a series are independent of each other, then the variance of the original series is used as an estimate of the variance of the random element. That procedure was carried out in this study.

Besides the usual aggregation problems, there are problems of aggregation which are unique to the variate difference method. To be specific, because all variability measures are derived from state time series data, it can be shown that the estimated variance of the random element for crop yields is a lower bound estimate of the true variability faced by an individual farmer. Carter and Dean (1960, p. 216) showed that the estimated random variance of the yield of the aggregated series for a particular crop is a function of: (1) the random yield correlation coefficient between individual farms producing that crop and (2) the

number of farms making up the aggregated series. If there were perfect
random yield correlation between farms (which is quite unrealistic), then
the estimated random variance of the aggregated series would be equal to an
estimated random variance derived from individual farm data. However, in
the more realistic case where there is not perfect yield correlation of
the random elements between farms, then the estimated variance from the
aggregate data varies inversely with the number of farms making up the
aggregate. The corresponding correlation with respect to price is very
nearly one, so the number of farms making up the aggregate price data
would not have a significant effect on the estimated random price variance
of the aggregated series.

One further limitation of the variate difference method exists.
The use of the method for estimating the random income variance leads to
internal inconsistencies, because income (per acre) is equal to the
product of price and yield (per acre). Both price and yield series consist of a mathematical expectation (systematic component) and a random
element, therefore, it becomes obvious that the income series (the product
of price and yield) will consist of error and nonerror products. Because
of the products of systematic and random terms, finite differencing can
never eliminate completely the mathematical expectation from the random
(error) terms. Our estimated random income variance is, therefore, an
upper bound approximation of the true random variability (Carter and
Dean, 1960, p. 218).

Absolute and Relative Variability
Measures

Farmers, in general, are interested in whether prices, yields,
and incomes associated with crops fluctuate a great deal. A statistical

measure of dispersion of a series will provide an estimate of the amount
of fluctuation. This type of estimate is a measure of absolute variability.
More important to a farmer pondering a decision to plant one crop or
another, or some combination of crops, or what livestock to raise is the
variability of one crop or species of livestock relative to others. A
statistical measure which takes into account different dispersions,
average values, and units will provide this information. We will refer to
this as a measure of relative variability.

In this study, the portion of the total variance of prices, yields,
and gross income that is "unpredictable" from the individual farmer's
standpoint is the estimated standard deviation determined by the variate
difference method. This is our empirical measure of absolute variability.

Equation (1) defines the "random variability coefficient."

$$(1)\ \text{Random Variability Coefficient} = \frac{\sqrt{\text{random variance}}}{\text{mean}} \times 100.$$

It is used as a measure of the variability of one crop (or species of
livestock) relative to others. It will be referred to as our empirical
measure of relative variability. Thus, relative variability is measured
by the standard deviation (determined by the variate difference method)
as a percentage of the mean level for each series under study.

The justification for the use of the above measures of absolute
and relative variability is straightforward.[1] The standard deviation,
instead of the variance of the random component of each series, is used as
a measure of absolute variability so as to have our measure reported in
the same unit as the original series. The variance of a series is
reported in squared units; whereas, the standard deviation is simply the
square root of the variance.

Because the standard deviation as a measure of dispersion is expressed in terms of the units of the variate, it is impossible to compare dispersions in different populations unless they happen to be measured in the same units. Also, as the size of the average value of a series can influence the magnitude of an absolute measure of dispersion, a measure of dispersion is needed which takes into account different mean values. Thus, measuring variability by using the standard deviation as a percentage of the mean value corrects for both problems and allows one to make variability comparisons among different populations with different mean values.

Results

The empirical estimates of absolute and relative variability of prices, yields, incomes, livestock numbers, and value are based on annual state observations for the period 1866 through 1909, except where otherwise noted for certain crops or livestock with shorter recorded histories. These estimates are computed for forty-eight states (Alaska and Hawaii are, of course, excluded). The data used for the computations are the revised statistics of the United States Department of Agriculture, whose sources are listed in Appendix A. A detailed discussion of the data (and their limitations); a listing of various incomplete series; and a discussion of several adjustments made to the data also are included in Appendix A.

The United States Department of Agriculture has farm data beginning in 1866 for eleven major crops--wheat, corn, oats, barley, rye, buckwheat, potatoes, sweet potatoes, tame hay, cotton, and tobacco--and for six species of livestock--horses, mules, milk cows, all cattle, sheep,

and hogs. This study contains empirical variability estimates for all of
these crops and livestock for the states in which the data exist. These
estimates are presented in Tables A1 through A144 in Appendix A. However,
before the specific results of the variability measures are discussed, it
is important to know how well the variate difference method works for this
particular set of data.

Upon examining the original results of finite differencing and
using historical information, it was our conclusion that this method of
detrending provided a quite reasonable estimate of the random component
of each time series.[2] The long-term decline in farm prices until 1896
and the increase in prices after 1896 were eliminated successfully from
the estimate of the variability of the random element of each price
series. Given that the trend of average yields of the major crops from
1866 through 1909 was nearly constant, one would expect that in many cases
the original series would be random. That was precisely what the results
showed for those series whose trends appeared constant from 1866 through
1909.

Given large changes in the numbers of livestock on farms in many
states and various cycles in the numbers during the late nineteenth
century, a sizeable systematic component and a smaller random component
were to be expected in the original series. Our results for the
livestock series showed that the random variability was a small proportion of total variability. Thus, we concluded that the detrending technique employed here worked well for the livestock series. It is interesting to note that livestock values (which had trends similar to crop
prices, in addition to their own cyclical patterns) also contained a random variability whose proportion of total variability was quite small.

Again, it appeared that the detrending technique worked well. Because
we were interested only in the "unpredictable" part of each time series,
it was encouraging that the method chosen for the purpose of estimating
random variability successfully eliminated systematic (predictable)
movements in each series.

General Observations

A few general observations about the results of the study will be
discussed in this section. These observations will include a comparison of
the crop estimates with the livestock estimates and a discussion of
regional patterns in the variability measures. Because many of the series
under study are measured in different units and practically all have
different mean values, our measure of relative variability will be used as
the basis for comparisons between different series. In later sub-sections,
the estimates of selected major crops will be discussed.

Price and Value Per Head Variability. When the variability measures
of price and value per head series are considered, some interesting conclusions can be drawn. As one might expect, the price variability of crops was
greater than the price variability of livestock. In fact, inspection of
Table 3-1 shows that, with few exceptions, the lowest price variability for
a crop was greater than the highest value per head variability.[3] The few
exceptions were the variability of the value of hogs for more than half of
the states (particularly in the "corn and hog belt" of the United
States); the variability of some livestock values in the Northeast
(Maine, Vermont, Massachusetts, Rhode Island, and Connecticut), and price
variability for many species of livestock for the less settled western
states (Montana, Arizona, Wyoming, Idaho, and New Mexico).

TABLE 3-1

SELECTED CROPS AND LIVESTOCK, REPRESENTATIVE SAMPLE OF STATES:
RANKING BY PRICE AND VALUE PER HEAD RANDOM VARIABILITY
COEFFICIENTS

MAINE Random Variability Coefficient		VERMONT Random Variability Coefficient		NEW YORK Random Variability Coefficient	
(Percent)		(Percent)		(Percent)	
Crop		**Crop**		**Crop**	
Potatoes	24.79	Potatoes	26.16	Potatoes	28.83
Tame Hay	14.41	Barley	10.67	Tobacco	18.24
Oats	10.68	Tame Hay	10.29	Oats	15.38
Buckwheat	8.94	Oats	9.64	Barley	11.60
Rye	8.57	Wheat	9.56	Buckwheat	11.13
Corn	7.75	Rye	9.22	Wheat	11.11
Wheat	7.20	Corn	9.04	Tame Hay	10.23
Barley	4.69	Buckwheat	6.43	Corn	9.24
				Rye	7.36
Livestock		**Livestock**		**Livestock**	
Hogs	10.33	Hogs	11.78	Hogs	7.07
All Cattle	8.69	Sheep	9.61	Milk Cows	6.19
Milk Cows	7.96	All Cattle	6.58	All Cattle	5.89
Sheep	7.51	Horses	5.11	Sheep	5.82
Horses	2.42	Milk Cows	4.68	Horses	3.10
				Mules	3.01

TABLE 3-1
(Continued)

OHIO Random Variability Coefficient		ILLINOIS Random Variability Coefficient		MINNESOTA Random Variability Coefficient	
(Percent)		(Percent)		(Percent)	
Crop		Crop		Crop	
Potatoes	35.43	Potatoes	39.35	Potatoes	43.22
Tobacco	22.37	Corn	30.25	Barley	18.66
Corn	18.16	Oats	28.08	Wheat	16.27
Tame Hay	13.62	Tobacco	22.88	Oats	16.19
Wheat	13.25	Tame Hay	19.86	Rye	13.59
Oats	12.99	Sweet		Tame Hay	13.11
Barley	12.21	Potatoes	16.09	Corn	11.33
Rye	10.85	Wheat	15.03	Buckwheat	10.61
Buckwheat	9.36	Barley	13.86		
		Buckwheat	12.31		
		Rye	12.19		
Livestock		Livestock		Livestock	
Hogs	9.36	Hogs	10.43	Mules	10.69
Sheep	5.16	Sheep	5.94	Hogs	9.50
All Cattle	3.34	All Cattle	4.20	Sheep	4.98
Mules	2.57	Milk Cows	3.73	All Cattle	4.45
Horses	2.17	Horses	3.64	Milk Cows	3.06
Milk Cows	1.81	Mules	3.41	Horses	3.04

TABLE 3-1
(Continued)

IOWA		NEBRASKA		KANSAS	
Random Variability Coefficient		Random Variability Coefficient		Random Variability Coefficient	
(Percent)		(Percent)		(Percent)	
Crop		Crop		Crop	
Potatoes	46.50	Potatoes	59.96	Corn	49.08
Corn	33.03	Corn	48.83	Potatoes	40.58
Barley	21.87	Oats	27.54	Tame Hay	27.84
Oats	21.08	Tame Hay	24.68	Oats	25.05
Rye	20.57	Rye	21.78	Sweet Potatoes	19.93
Wheat	18.39	Wheat	20.38	Wheat	19.75
Sweet Potatoes	17.06	Barley	17.80	Buckwheat	17.77
Buckwheat	11.60	Buckwheat	10.90	Rye	17.54
Tame Hay	10.13			Barley	14.67
Livestock		Livestock		Livestock	
Hogs	9.87	Hogs	14.38	Hogs	16.63
All Cattle	5.03	Mules	7.21	Sheep	7.30
Milk Cows	4.74	Sheep	5.23	All Cattle	5.39
Mules	4.22	All Cattle	5.09	Milk Cows	4.26
Sheep	4.03	Horses	5.00	Mules	3.28
Horses	3.38	Milk Cows	4.92	Horses	2.75

TABLE 3-1
(Continued)

NORTH CAROLINA Random Variability Coefficient		GEORGIA Random Variability Coefficient		TENNESSEE Random Variability Coefficient	
(Percent)		(Percent)		(Percent)	
Crop		Crop		Crop	
Barley	18.63	Corn	12.04	Rye	26.43
Tobacco	15.40	Potatoes	11.01	Corn	25.58
Corn	11.39	Oats	10.41	Potatoes	23.67
Sweet		Sweet		Tobacco	23.12
Potatoes	10.59	Potatoes	9.76	Sweet	
Tame Hay	10.52	Tame Hay	9.55	Potatoes	18.74
Wheat	10.16	Wheat	8.89	Barley	14.68
Cotton	9.63	Rye	7.86	Wheat	13.25
Oats	9.47	Cotton	7.77	Oats	12.20
Potatoes	9.33	Barley	7.63	Buckwheat	9.51
Buckwheat	7.90			Cotton	9.49
Rye	6.37			Tame Hay	9.44
Livestock		Livestock		Livestock	
Milk Cows	9.57	Hogs	7.72	Hogs	12.28
Sheep	6.69	All Cattle	6.40	All Cattle	6.32
Hogs	6.22	Milk Cows	5.45	Milk Cows	5.46
Mules	5.67	Horses	5.19	Sheep	3.54
All Cattle	4.94	Mules	4.81	Horses	3.51
Horses	3.68	Sheep	4.32	Mules	2.90

TABLE 3-1
(Continued)

ALABAMA		ARKANSAS		TEXAS	
Random Variability Coefficient		Random Variability Coefficient		Random Variability Coefficient	
(Percent)		(Percent)		(Percent)	

Crop		Crop		Crop	
Corn	12.57	Potatoes	26.02	Corn	32.93
Wheat	10.54	Corn	22.20	Barley	25.21
Tame Hay	10.35	Tobacco	18.94	Sweet	
Rye	10.15	Rye	14.88	Potatoes	20.32
Cotton	9.99	Wheat	13.10	Tame Hay	19.27
Oats	8.77	Sweet		Oats	16.79
Potatoes	8.58	Potatoes	12.46	Potatoes	16.00
Barley	8.44	Oats	12.19	Rye	15.26
Sweet		Tame Hay	10.42	Wheat	15.05
Potatoes	8.38	Cotton	10.04	Cotton	10.45

Livestock		Livestock		Livestock	
Hogs	9.81	Hogs	16.54	Hogs	8.65
Sheep	7.46	Sheep	8.32	All Cattle	7.03
Horses	6.29	Horses	6.39	Sheep	6.27
All Cattle	6.15	Mules	5.75	Mules	5.84
Milk Cows	5.28	Milk Cows	4.79	Milk Cows	4.87
Mules	3.92	All Cattle	4.42	Horses	3.96

TABLE 3-1
(Continued)

MONTANA		UTAH		CALIFORNIA	
Random Variability Coefficient		Random Variability Coefficient		Random Variability Coefficient	
(Percent)		(Percent)		(Percent)	
Crop		Crop		Crop	
Potatoes	25.44	Potatoes	31.59	Potatoes	19.10
Barley	19.16	Rye	18.28	Barley	17.53
Tame Hay	18.85	Tame Hay	11.39	Sweet	
Oats	12.53	Oats	11.04	Potatoes	16.74
Wheat	9.41	Corn	10.30	Tame Hay	16.33
Rye	7.11	Barley	8.61	Rye	14.02
		Wheat	7.99	Wheat	12.99
				Corn	12.32
				Oats	9.07
Livestock		Livestock		Livestock	
Hogs	17.51	Hogs	10.90	Hogs	12.37
Sheep	8.30	Mules	7.86	Sheep	7.46
All Cattle	7.76	Milk Cows	7.50	Mules	5.76
Milk Cows	7.72	All Cattle	6.95	All Cattle	5.61
Horses	7.06	Sheep	5.39	Horses	4.20
Mules	4.93	Horses	4.68	Milk Cows	3.84

Source: See the introduction to Appendix A. For any qualifications to the individual estimates, see the notes to the corresponding tables in Appendix A.

It is reasonable to suspect that crops are more susceptible than livestock to the random occurrences which affect both crop and livestock production--droughts, floods, rainstorms, prairie fires, and insects. Thus, random fluctuations in the supply of crops to the market are greater than the unpredictable shifts in the total stock of livestock. On the demand side, the factors that cause random fluctuations--domestic employment conditions, foreign tariffs and other restrictions, and changes of supply (which affect the level of export demand) in foreign producing nations and importing countries--probably affect crops and livestock alike. It appears reasonable (for the above reasons) that the measured random price variability of crops would be greater than the estimates of random value per head variability of livestock.

The differences in the magnitude of the random price variability estimates between crops and livestock can be viewed from another perspective. A simple stock-flow analysis will help to explain why price variability of crops is greater than price variability of livestock. The greater part of the total stock of a crop in existence is produced from year to year. Very little is carried over from a previous season. Thus, any random fluctuations in the flow of production will cause a corresponding fluctuation in the total stock of the crop available for that season. Because the flow of production of livestock for the market is quite small compared to the total stock in existence, the carry-over from one year to the next is a large proportion of the total stock. When a random occurrence affects livestock production, it has a negligible effect on the total amount of livestock. The random variability of of value per head also is not affected as much as if the ratio of carry-over to total stock were quite small.

Yield and Livestock Numbers Variability. A comparison of random yield variability and the random variability of livestock numbers on farms for the postbellum period leads to an expected conclusion. Except for the variability associated with the number of mules on farms for many Western states (Montana, Idaho, Wyoming, Arizona, Utah, Nevada, and Washington), the lowest random yield variability in each state was greater than the highest measured variability of livestock numbers. The analysis which explained the differences in price variability between crops and livestock would lead one to expect similar differences between crop yields and livestock numbers. The total stock of animals on farms is simply so much larger compared to the number that may be affected by random factors (which affect both crops and livestock) that the "unpredictable" variability of livestock numbers is small compared to the random yield variability of crops. Variability measures of crop yields and livestock numbers for a representative sample of states are presented in Table 3-2.[4]

Gross Income Variability. The random variability measures of income per acre did show some consistent patterns when they were compared to price and yield variability estimates. If Tables 3-1 and 3-2 are compared with the income variability coefficients presented in Table 3-3, it appears that, in general, the crops with the highest price and yield variability coefficients had relatively high income variability, and crops with low price and yield variability usually had low income variability. However, there were specific crops for which neither statement was an accurate reflection of the relationship among price, yield, and income random variability coefficients.

Income per acre for a crop is simply the product of price and yield per acre. Thus, the income variability coefficients are a function

TABLE 3-2

SELECTED CROPS AND LIVESTOCK, REPRESENTATIVE SAMPLE OF STATES:
RANKING BY YIELD AND NUMBER ON FARMS RANDOM VARIABILITY
COEFFICIENTS

MAINE Random Variability Coefficient		VERMONT Random Variability Coefficient		NEW YORK Random Variability Coefficient	
(Percent)		(Percent)		(Percent)	
Crop		Crop		Crop	
Potatoes	20.07	Potatoes	16.19	Potatoes	17.08
Rye	15.32	Buckwheat	12.76	Wheat	16.55
Corn	15.16	Corn	12.40	Oats	12.87
Oats	14.28	Rye	11.53	Buckwheat	12.67
Tame Hay	12.10	Wheat	9.29	Barley	12.11
Buckwheat	11.43	Barley	8.21	Tame Hay	10.72
Wheat	8.78	Tame Hay	7.84	Tobacco	10.23
Barley	3.67	Oats	7.41	Rye	9.39
				Corn	9.08
Livestock		Livestock		Livestock	
Sheep	1.85	Sheep	1.36	Mules	4.48
Milk Cows	1.35	Hogs	1.21	Sheep	1.78
Hogs	1.33	All Cattle	.63	Hogs	1.34
All Cattle	1.03	Milk Cows	.56	All Cattle	.81
Horses	.67	Horses	0.33	Horses	0.28
				Milk Cows	0.17

TABLE 3-2
(Continued)

OHIO Random Variability Coefficient		ILLINOIS Random Variability Coefficient		MINNESOTA Random Variability Coefficient	
(Percent)		(Percent)		(Percent)	
Crop		Crop		Crop	
Wheat	22.82	Potatoes	27.76	Potatoes	16.78
Potatoes	17.51	Wheat	21.64	Wheat	15.88
Rye	14.35	Corn	19.61	Buckwheat	15.73
Tobacco	14.14	Sweet		Corn	13.55
Buckwheat	13.34	Potatoes	18.49	Oats	13.50
Oats	12.98	Buckwheat	17.17	Tame Hay	11.87
Corn	12.17	Oats	16.43	Barley	7.86
Tame Hay	11.99	Tame Hay	15.21	Rye	5.34
Barley	10.49	Tobacco	9.36		
		Barley	8.73		
		Rye	4.60		
Livestock		Livestock		Livestock	
Sheep	2.10	Hogs	3.46	Mules	4.03
Hogs	2.01	Sheep	2.56	Hogs	3.48
Mules	1.52	All Cattle	1.61	Sheep	3.01
All Cattle	.83	Mules	.99	All Cattle	1.36
Milk Cows	.36	Milk Cows	.75	Milk Cows	1.24
Horses	.35	Horses	.35	Horses	1.09

TABLE 3-2
(Continued)

IOWA		NEBRASKA		KANSAS	
	Random Variability Coefficient		Random Variability Coefficient		Random Variability Coefficient
	(Percent)		(Percent)		(Percent)
Crop		**Crop**		**Crop**	
Potatoes	21.54	Corn	25.98	Barley	35.77
Sweet		Potatoes	24.70	Potatoes	34.27
Potatoes	16.64	Barley	23.12	Corn	31.07
Corn	15.96	Oats	19.95	Wheat	22.30
Oats	14.99	Buckwheat	16.86	Tame Hay	20.09
Wheat	14.08	Tame Hay	15.76	Sweet	
Buckwheat	12.48	Rye	13.17	Potatoes	19.70
Tame Hay	11.98	Wheat	10.68	Oats	16.78
Barley	9.09			Rye	13.94
Rye	6.89				
Livestock		**Livestock**		**Livestock**	
Sheep	3.86	Hogs	10.52	Hogs	5.23
Hogs	3.05	Sheep	6.31	Sheep	4.79
All Cattle	2.16	All Cattle	2.86	Horses	1.64
Mules	1.96	Horses	1.98	All Cattle	1.43
Horses	1.02	Mules	1.62	Mules	1.38
Milk Cows	.94	Milk Cows	1.39	Milk Cows	.87

TABLE 3-2
(Continued)

NORTH CAROLINA Random Variability Coefficient		GEORGIA Random Variability Coefficient		TENNESSEE Random Variability Coefficient	
(Percent)		(Percent)		(Percent)	
Crop		Crop		Crop	
Rye	20.67	Wheat	20.38	Wheat	25.96
Wheat	20.67	Rye	17.65	Sweet	
Oats	15.68	Sweet		Potatoes	18.29
Cotton	12.84	Potatoes	14.12	Tobacco	17.48
Corn	12.53	Corn	13.01	Cotton	16.80
Buckwheat	11.33	Cotton	12.98	Oats	15.33
Sweet		Oats	10.95	Barley	14.44
Potatoes	11.18	Tobacco	10.27	Rye	13.56
Tame Hay	11.11	Potatoes	8.66	Buckwheat	13.21
Tobacco	9.99	Tame Hay	8.12	Corn	11.60
Potatoes	8.78			Potatoes	10.88
				Tame Hay	6.04
Livestock		Livestock		Livestock	
Hogs	2.85	Hogs	1.48	Hogs	1.73
Sheep	1.56	Sheep	1.21	Milk Cows	1.41
All Cattle	1.09	All Cattle	.75	Sheep	.81
Mules	.80	Milk Cows	.58	Mules	.53
Horses	.76	Mules	.56	All Cattle	.37
Milk Cows	.64	Horses	.36	Horses	.30

TABLE 3-2
(Continued)

ALABAMA Random Variability Coefficient		ARKANSAS Random Variability Coefficient		TEXAS Random Variability Coefficient	
(Percent)		(Percent)		(Percent)	
Crop		**Crop**		**Crop**	
Wheat	20.56	Sweet		Sweet	
Cotton	15.53	Potatoes	27.67	Potatoes	24.67
Sweet		Wheat	20.87	Wheat	24.57
Potatoes	14.95	Cotton	17.54	Barley	21.47
Oats	12.53	Corn	14.46	Corn	21.07
Corn	12.12	Rye	14.34	Cotton	20.78
Tame Hay	8.72	Oats	13.73	Oats	17.19
Potatoes	7.65	Potatoes	13.14	Rye	16.11
		Tame Hay	10.84	Tame Hay	12.89
		Tobacco	9.50	Potatoes	12.68
Livestock		**Livestock**		**Livestock**	
Hogs	1.82	Hogs	5.71	Hogs	4.06
Milk Cows	.98	All Cattle	1.82	Sheep	3.23
Sheep	.78	Mules	1.44	Milk Cows	2.49
All Cattle	.69	Sheep	1.33	All Cattle	2.09
Mules	.61	Milk Cows	1.13	Mules	1.20
Horses	.43	Horses	1.12	Horses	.56

TABLE 3-2
(Continued)

MONTANA Random Variability Coefficient		UTAH Random Variability Coefficient		CALIFORNIA Random Variability Coefficient	
(Percent)		(Percent)		(Percent)	
Crop		Crop		Crop	
Rye	19.14	Rye	16.80	Rye	23.32
Corn	16.19	Wheat	13.74	Barley	17.94
Barley	14.72	Potatoes	12.79	Wheat	16.01
Wheat	14.33	Corn	10.50	Oats	12.71
Potatoes	12.04	Barley	8.66	Sweet	
Tame Hay	9.27	Tame Hay	6.62	Potatoes	11.45
Oats	8.16	Oats	4.87	Potatoes	10.81
				Corn	10.47
				Tame Hay	7.88
Livestock		Livestock		Livestock	
Mules	10.36	Mules	7.21	Sheep	3.33
Sheep	3.76	Sheep	3.65	All Cattle	1.22
Milk Cows	2.87	Hogs	1.37	Hogs	1.19
All Cattle	2.65	All Cattle	1.23	Milk Cows	.94
Hogs	2.63	Milk Cows	.89	Mules	.81
Horses	1.15	Horses	.33	Horses	.50

Source: See the introduction to Appendix A. For any qualifications to individual estimates, see the notes to the corresponding tables in Appendix A.

·TABLE 3-3

SELECTED CROPS, REPRESENTATIVE SAMPLE OF STATES: RANKING BY
GROSS INCOME PER ACRE RANDOM VARIABILITY
COEFFICIENTS

MAINE Random Variability Coefficient		VERMONT Random Variability Coefficient		NEW YORK Random Variability Coefficient	
(Percent)		(Percent)		(Percent)	
Crop		Crop		Crop	
Potatoes	19.15	Potatoes	21.11	Tobacco	26.38
Corn	16.81	Rye	15.83	Potatoes	24.43
Rye	14.85	Barley	13.09	Barley	19.27
Buckwheat	12.13	Buckwheat	12.06	Wheat	16.89
Oats	12.02	Corn	11.65	Oats	12.73
Wheat	9.56	Wheat	10.51	Buckwheat	11.56
Tame Hay	9.54	Oats	9.89	Corn	10.07
Barley	4.99	Tame Hay	8.62	Rye	9.36
				Tame Hay	6.54

OHIO Random Variability Coefficient		ILLINOIS Random Variability Coefficient		MINNESOTA Random Variability Coefficient	
(Percent)		(Percent)		(Percent)	
Crop		Crop		Crop	
Tobacco	28.72	Tobacco	25.98	Wheat	28.62
Potatoes	27.30	Wheat	20.50	Potatoes	25.68
Wheat	16.58	Oats	16.22	Oats	17.04
Rye	16.54	Corn	15.11	Rye	14.73
Buckwheat	13.60	Barley	14.82	Buckwheat	13.62
Barley	12.94	Sweet Potatoes	14.19	Barley	12.29
Corn	11.44	Buckwheat	13.42	Corn	10.43
Oats	10.35	Rye	11.94	Tame Hay	9.85
Tame Hay	8.11	Tame Hay	11.11		
		Potatoes	10.94		

TABLE 3-3
(Continued)

IOWA Random Variability Coefficient		NEBRASKA Random Variability Coefficient		KANSAS Random Variability Coefficient	
(Percent)		(Percent)		(Percent)	
Crop		Crop		Crop	
Potatoes	35.34	Potatoes	26.39	Barley	26.42
Sweet		Rye	22.94	Potatoes	23.43
Potatoes	23.60	Barley	22.25	Rye	18.84
Wheat	19.93	Wheat	18.38	Wheat	15.32
Tame Hay	19.10	Oats	17.86	Oats	15.01
Rye	17.59	Buckwheat	15.04	Sweet	
Barley	16.66	Corn	14.80	Potatoes	13.57
Oats	14.14	Tame Hay	14.10	Corn	13.15
Corn	9.55			Tame Hay	7.06
Buckwheat	6.60				

NORTH CAROLINA Random Variability Coefficient		GEORGIA Random Variability Coefficient		TENNESSEE Random Variability Coefficient	
(Percent)		(Percent)		(Percent)	
Crop		Crop		Crop	
Wheat	18.57	Wheat	34.05	Wheat	35.24
Tobacco	17.81	Rye	25.35	Rye	26.51
Rye	17.50	Sweet		Tobacco	23.06
Cotton	16.98	Potatoes	17.02	Barley	20.01
Oats	16.03	Tobacco	15.29	Sweet	
Buckwheat	13.46	Potatoes	14.92	Potatoes	18.23
Sweet		Oats	14.37	Buckwheat	15.21
Potatoes	12.61	Corn	12.17	Cotton	14.66
Potatoes	11.88	Cotton	11.50	Oats	13.97
Tame Hay	11.76	Tame Hay	8.11	Potatoes	12.54
Corn	10.48			Tame Hay	11.26
				Corn	8.53

TABLE 3-3
(Continued)

ALABAMA Random Variability Coefficient		ARKANSAS Random Variability Coefficient		TEXAS Random Variability Coefficient	
(Percent)		(Percent)		(Percent)	
Crop		Crop		Crop	
Wheat	19.11	Wheat	25.67	Barley	29.41
Tame Hay	13.75	Tobacco	20.43	Wheat	24.43
Sweet		Sweet		Sweet	
Potatoes	13.19	Potatoes	19.98	Potatoes	22.01
Corn	12.30	Rye	18.58	Potatoes	19.94
Oats	12.29	Cotton	17.94	Tame Hay	17.69
Cotton	11.73	Potatoes	16.63	Cotton	17.66
Potatoes	9.06	Oats	13.80	Rye	17.17
		Corn	13.01	Oats	15.03
		Tame Hay	10.94	Corn	10.14

MONTANA Random Variability Coefficient		UTAH Random Variability Coefficient		CALIFORNIA Random Variability Coefficient	
(Percent)		(Percent)		(Percent)	
Crop		Crop		Crop	
Wheat	27.74	Rye	28.85	Rye	22.40
Potatoes	24.91	Wheat	23.59	Barley	17.88
Corn	19.83	Potatoes	21.82	Wheat	17.14
Tame Hay	17.63	Tame Hay	20.10	Potatoes	16.21
Barley	17.38	Corn	13.71	Oats	15.70
Rye	16.28	Barley	12.35	Corn	13.85
Oats	12.14	Oats	10.34	Tame Hay	13.69
				Sweet	
				Potatoes	13.34

Source: See the introduction to Appendix A. For any qualifications to individual estimates, see the notes to the corresponding tables in Appendix A.

of price and yield random variability coefficients. Also, the relationship between year-to-year changes in both prices and yields can be
paramount in determining the relationship among price, yield, and
income variability estimates. If the correlation between price and
yield from one year to the next is negative, as is usually suspected, then
this has a tendency to reduce income variability. However, if there is a
positive correlation between year-to-year changes in price and yields--low
prices tend to be associated with low yields and vice versa--then income
variability will be larger. The random swings in income are accentuated
by this positive correlation.

This latter phenomenon became prevalent for some crops during the
nineteenth century as markets expanded in size from local to national and
eventually to international. Thus, market prices became dependent not only
on local conditions but on international conditions. A farmer in
Minnesota who had a low yield one year might receive a very low price because
of a bumper crop in Australia or a high price because of a very bad crop
in Australia. With prices determined in world markets, a farmer in the
United States might very well face increased income variability, because
of the possibilities of positive correlation between year-to-year changes
in price and yields.

Regional Differences in Price and Value Per Head Variability.
Important regional differences exist in the variability estimates associated with the major agricultural sectors of the United States. Upon inspection of Tables 3-1 through 3-3, one becomes aware of the obvious patterns.

The regional differences in random price variability can be interpreted from the sample of states in Table 3-1. As might be expected, prices
were more variable in the less settled and newer agricultural regions of the

country; whereas, the least variable of all regions were the older areas of the South and of the Northeast. For the census regions, which correspond to major agricultural sectors of the United States, the ranges of the random price variability coefficients associated with crops were as follows: South Atlantic states, 27 percent to 7 percent; North Atlantic states, 29 percent to 5 percent; South Central states, 33 percent to 8 percent; East North Central states, 39 percent to 8.5 percent; and West North Central states, 60 percent to 10 percent. However, the regional differences in the estimates were even more striking than was indicated by the preceding ranges.

A close inspection of the tables presented in Appendix A shows more obvious regional patterns. For the "old" South, the range of price estimates was concentrated between 13 percent and 8 percent, while the newer areas (for example, Arkansas and Texas) increased the range to as high as 33 percent. The price variability estimates for the North Atlantic states actually were skewed toward the lower end of the range. In fact, the greater part of the variability coefficients—about 80 percent of them—measured less than 12 percent.

While the East North Central states' range was from 39 percent to 8.5 percent, nearly all of the estimates were well distributed within a 20 percent to 10 percent range. For the West North Central region, the state variability estimates were rarely more than 40 percent. However, the greater part of the estimates—about 75 percent of them—were more than 15 percent and many were more than 20 percent.

The measures of the random variability of livestock value per head did not show as much regional difference as crop prices, with one possible exception. The variability of hog prices, as can be seen in Table 3-1, was higher for the "corn and hog" belt of the United States than for most other areas of the country. All other variability measures of livestock prices were in approximately the same range across the country.

What accounted for these regional differences? Because crop prices are determined by the demand for and supply of crops, it must be that certain factors affecting random fluctuations in demand and supply were different for different regions of the United States. Greater random variabilities on the supply side would be expected for new areas. Farmers in a new area would have had to familiarize themselves with different climatic conditions, new soils and terrain, different types of insects, and new techniques of production. These factors accounted for most of the difference between the states of the Northeast and South and the states of the North Central region. The difference between the East and West North Central states, as well as the difference between the western and eastern parts of the South Central states, could be explained, in great part, in this way. Nevertheless, the demand for crops also affected crop price variability.

Did demand have different random fluctuations between regions? The older agricultural regions of the United States—the North Atlantic, South Atlantic, and parts of the South Central and East North Central states— appeared to have had more established and steadier markets for their crops than did new areas. That would account for some difference in demand fluctuations. Also, as mentioned previously in this chapter, the demand for some products (wheat and pork) depended upon the supply conditions and political constraints in other countries. Export markets were quite susceptible to crop failures and to tariff restrictions imposed elsewhere, without much warning to American producers. Because foreign competition was relatively not large in cotton, this last factor did not seriously affect the major cash crop of the South. Yet the grain crops of the West North Central states definitely were affected by it.

The different conditions on the supply side of crops and the factors mentioned as affecting demand combined to make prices more uncertain as one moved from east to west or from older to new areas.

Regional Differences in Yield and Number on Farms Variability. The pattern of regional differences in the estimates of random yield variability is represented by the sample of states in Table 3-2. The differences were not quite as striking as the regional differences in random price variability. Nevertheless, distinctions can be made. There also were definite distinctions on a regional basis for the random variability associated with the number of livestock on farms. These differences were more striking than those pertaining to the random variability associated with the value per head of livestock.

For the census regions which correspond to major agricultural sectors of the United States, the ranges of the random yield variability coefficients associated with the crop sample were as follows: South Atlantic states, 23 percent to 8 percent; North Atlantic states, 25 percent to 4 percent; South Central states, 28 percent to 5 percent; East North Central states, 28 percent to 8 percent; and West North Central states, 36 percent to 7 percent. For the same census regions, the ranges of the random number-on-farms variability coefficients associated with the livestock sample were as follows: South Atlantic, North Atlantic, and South Central states, 3 percent to .2 percent; East North Central states, 5.5 percent to .33 percent; and West North Central states, 10.5 percent to

1 percent. However, the regional patterns actually were more pronounced
than was indicated by the preceding information.

A careful study of all states belonging to the North Atlantic region
indicates that only six out of a total of seventy-three yield estimates
were greater than the 17 percent variability range.[5] For all states in the
South Atlantic region, only sixteen out of seventy-two estimates were more
than the 17 percent variability range. The South Central states possessed a
level of yield variability similar to the South Atlantic states, except for
the western parts of the region (for example, Arkansas and Texas). Those
western states contained most of the yield variability estimates which were
more than 18 percent and nearly all estimates more than 20 percent.

The region of the United States with the highest degree of random
yield variability was the North Central. The East North Central states
displayed characteristics similar to those of the South Central region, while
the West North Central states' variability estimates were quite different
from the estimates of all other regions. Not only was the range 36 percent to 7 percent, but many estimates were skewed toward the higher end.
A sample of three states from this region--Kansas, North Dakota, and South
Dakota--makes the point quite clear.[6] For the three states, seventeen out
of a total of twenty-two random yield variability coefficients were more
than 19 percent, and several were in the 30 percent range.

The explanation for these divergent regional patterns should be
clear. Again, the West North Central states had by far the most random
variability. This pattern was due to farmers moving into new areas with
which they had little (or at least, less) knowledge about climatic conditions, soil and terrain types, insect problems, and techniques of production best suited to the area. Also, unpredictable technological change

throughout this period might well account for part of the fluctuations
in crop yields (Parker, 1972, pp. 381-391).

Regional Differences in Income Variability. Regional patterns of
income variability per acre were not nearly as pronounced as the regional
differences in price and yield variability. The ranges of the random
income variability coefficients for the census regions were as follows:
North Atlantic states, 29 percent to 5 percent; East North Central states,
35 percent to 8 percent; West North Central states, 40 percent to 7
percent; South Atlantic states, 35 percent to 8 percent; and South Central
states, 35 percent to 8 percent. Even though the preceding figures appear
to indicate uniformity among the regions, close inspection of the estimates
for states within regions shows some differences.

The sample of states presented in Table 3-3 gives some indication
of these differences. The greater part of the estimates of income variability per acre for states such as Iowa and Nebraska was more than 17 percent; whereas, the vast majority of estimates for Ohio and Illinois was
less than 17 percent. Income variability was quite high in North and South
Dakota. Nine coefficients out of a total of fourteen were more than 20
percent.[8] Although the range of income variability was identical in both
the South Atlantic and South Central regions, some intraregional difference
existed. For example, the greater part of income variability estimates
associated with the crop sample in Arkansas and Texas was more than 17
percent; whereas, the greater part of estimates from other states in the
South was less than 17 percent.

The border states in the South also had higher income variability
than the others. In the South Atlantic states, only thirteen out of a total

of seventy estimates were more than 20 percent. Lastly, in the North Atlantic region, fifty-five out of seventy-five income variability estimates were less than 15 percent, even though the range ran from 29 percent to 5 percent.

Again, there existed the same general regional differences with income variability as with price and yield variability. This was expected. Income variability, as mentioned earlier, is a function of both price and yield fluctuations plus the relationship between year-to-year changes in prices and yields. The correlation between prices and yields for some of the crops grown in the West North Central states was such that it should lead to higher levels of uncertain returns than for other regions.

Besides the effect on variability caused by these states being "frontier" areas, the grain producers' (and meat exporters') markets were expanding so that the demand these states faced now was dependent upon climatic conditions, tariff regulations, and quality standards (which could change without much warning) in foreign supplying areas and importing countries (Parker, 1972, pp. 405-406). The crops of the West North Central states did show evidence of positive correlation in the year-to-year relationship between prices and yields. Thus, this added to the amount of income variability facing the farmers.

From the previous discussion, we know that the preceding demand conditions did not hold for the cotton producers of the South, even though their markets were international in scope. That was because few foreign competitors produced cotton (Parker, 1972, p. 406). In fact, there is less evidence of year-to-year correlation between prices and yields for states in the South. Therefore, income uncertainty is primarily a function of the levels of random variability in prices and yields; whereas, the year-to-year relationship between prices and yields is not an important contributing factor.

If the year-to-year relationship between prices and yields tends to be negative, income variability is reduced. This applies to the North Atlantic states and states of the East North Central region where acreage was relatively constant from year-to-year during this period. Thus, total output was changed (and, hence, prices), primarily because of fluctuating yields. Therefore, with this negative correlation between year-to-year changes in prices and yields, there was even lower income variability for several crops than the individual levels of price and yield variability would have indicated.

The observations, in this section, on income variability are meant to be only general in nature. After discussing price and yield variability for the major crops of the postbellum period, we will deal specifically with the relationship between price variability, yield variability, the price-yield correlation, and income variability associated with the major crops. The discussion in the next three sections will be limited only to the major crops and the major producing states.

Random Price Variability of Major Crops

Discussed in this section is the price variability associated with five major crops—corn, cotton, hay, oats, and wheat—for the thirteen leading producing states at the end of the nineteenth century. The estimates of the random variability coefficients of prices are contained in Table 3-4. The crop with the highest variability was corn. Oats ranked next in price variability with wheat and hay (at about the same level) somewhat lower. Cotton easily had the lowest measured variability.

TABLE 3-4

FIVE MAJOR CROPS (AND HOGS), THIRTEEN LEADING PRODUCERS:
RANKING BY PRICE RANDOM VARIABILITY COEFFICIENTS

COTTON: Ranking by Price
Random Variability Coefficients

CORN: Ranking by Price
Random Variability Coefficients

State	Random Variability Coefficients (Percent)	State	Random Variability Coefficients (Percent)
Florida	26.23	Kansas	49.08
Oklahoma	17.78	Nebraska	48.83
Virginia	17.72	Missouri	39.93
South Carolina	11.63	Oklahoma	34.27
Texas	10.45	Iowa	33.03
Louisiana	10.13	Texas	32.93
Arkansas	10.04	Illinois	30.25
Alabama	9.99	Tennessee	25.58
North Carolina	9.63	Kentucky	24.61
Mississippi	9.51	Indiana	20.21
Tennessee	9.49	Ohio	18.61
Georgia	7.77	Georgia	12.04
		North Carolina	11.39

OATS: Ranking by Price
Random Variability Coefficients

WHEAT: Ranking by Price
Random Variability Coefficients

State	Random Variability Coefficients (Percent)	State	Random Variability Coefficients (Percent)
Illinois	28.08	North Dakota	24.43
Nebraska	27.54	South Dakota	24.09
Kansas	25.05	Nebraska	20.38
Iowa	21.08	Kansas	19.75
Missouri	19.48	Iowa	18.39
Wisconsin	16.88	Minnesota	16.27
Texas	16.79	Missouri	15.79
Michigan	16.67	Illinois	15.03
Minnesota	16.19	Indiana	13.88
New York	15.38	Michigan	13.46
Indiana	14.54	Ohio	13.25
Pennsylvania	14.40	California	12.99
Ohio	12.99	Pennsylvania	12.06

TABLE 3-4
(Continued)

State	Random Variability Coefficients (Percent)		Random Variability Coefficients (Percent)
HAY: Ranking by Price **Random Variability Coefficients**		**HOGS: Ranking by Price** **(Value Per Head)** **Random Variability Coefficients**	
Kansas	27.84	Kansas	16.63
Missouri	24.59	Arkansas	16.54
Wisconsin	24.42	Missouri	16.35
Illinois	19.86	Nebraska	14.38
California	16.33	Wisconsin	13.20
Indiana	15.82	Tennessee	12.28
Michigan	15.49	Kentucky	10.53
Ohio	13.62	Indiana	10.47
Minnesota	13.11	Illinois	10.43
Virginia	10.49	Iowa	9.87
New York	10.23	Ohio	9.36
Iowa	10.13	Texas	8.65
Pennsylvania	6.73	Georgia	7.72

Source: See the introduction to Appendix A. For any qualifications to the individual estimates, see the notes to the corresponding tables in Appendix A.

For ease in determining the ranking of price variability, the mean value of the random variability coefficients associated with each crop for the thirteen states were calculated as follows: corn, 29.5 percent; oats, 18.9 percent; wheat, 16.9 percent; hay, 16.0 percent; and cotton, 12.5 percent.

In general, Table 3-4 also supports the contention that crops were riskier as one moved from eastern states to western states or from older areas to newer areas. However, there were obvious exceptions. Illinois had a lot higher variability coefficients for hay and oats than did Iowa and Minnesota. In fact, surprisingly, Minnesota had relatively low variability measures for all of the major crops grown in the state, and Illinois had relatively high variability for all of its major crops, except wheat. Nevertheless, the general pattern still held.

The West North Central states ranked as the riskiest region in terms of prices associated with corn, hay, oats, and wheat. The North Atlantic region consistently ranked as the least variable area for the same crops along with most states of the "old" South. States of the "newer" South--for example, Arkansas and Texas--and the East North Central states usually ranked in the middle range of price variability. The ranking of price variability for cotton states was interesting. Except for Florida and Virginia, which were quite marginal producing areas, and Oklahoma, which was quite new, there were little differences in the region. Georgia was the only state that was very different from the other southern states, and it had quite low price variability.

The reasons for regional patterns of the variability coefficients were discussed previously. However, the differences between the major crops have not been analyzed. Why were corn prices a lot more variable than the prices of hay, oats, and wheat? Why did the price of cotton have a lot less random fluctuation than the prices of the other crops? The answers are not as clear as the questions. Nevertheless, some reasonable explanations account for the results.

One observation that pertains to all of the major crops is that the demand curve for each of them was inelastic in the relevant ranges.[9] This factor helps to explain why there was a high level of price variability, in general, for late nineteenth century agriculture. However, it cannot explain why some crop prices, in particular, were more variable than others. Crop prices will tend to fluctuate more for a given change in total production when farmers are facing price elasticities of demand that are less than unity, than if the quantity demanded is more responsive to price changes. Nonetheless, the demand for corn was estimated to be the least inelastic (and to have the greatest price variability); whereas, wheat was the most inelastic, with cotton being the crop second to wheat in terms of having the lowest estimated price elasticity of demand (Shultz, 1938; pp. 275-281, 320-330, 397-401). (Notice that Table 3-4 shows cotton to have the lowest price variability.)

The variability of the price of corn is probably the easiest to explain. The greater part of the United States' corn crop always has been used to feed hogs; hence, there is an intimate connection between the demand for and production of hogs and the demand for and production of corn. This creates an interconnection between the value of hogs and the price of corn. If the value of hogs rose, the demand for corn as feed increased and very little of the corn supply went to the market--most of the corn was fed to hogs on the same farm--and the price of corn then increased. Larger stocks of hogs placed downward pressure on hog prices, causing less demand for feed and more of the corn supply being placed on the market, which tended to lower corn prices. The previous rise of corn prices also had the effect of making hog production less profitable, thus adding to a lowered demand for feed. Those circumstances created the corn-hog cycle, which causes the price of corn (and the value of hogs) to have high levels of variability.[10] (Value per head variability of hogs is included in Table 3-4 as a point of reference.)

Why did cotton have the lowest estimated price random variability coefficients of the major crops? It cannot be explained by cotton's elasticity of demand, because the demand for cotton during this period was quite inelastic. That should have led to greater price fluctuations, not less. Some possible explanations follow: (1) changes in business conditions, which affected the level of demand for cotton, were such that demand for cotton tended to shift in the same direction as shifts in the supply of cotton. Thus, it could lead to the possibility of small fluctuations in price over time (Shultz, 1938, pp. 320-330); (2) because producers of cotton in the United States had little foreign competition, there were fewer outside factors (than for some other crops) to cause random fluctuation in its export demand (Shannon, 1945, pp. 110-112); and (3) there were well-established markets (cotton exchanges) for the crop, which may have had a stabilizing influence on its price (Shannon, 1945, pp. 112-117).

How can the price variability of oats, wheat, and hay be explained? Their ranking was in the order given, while corn was more variable and cotton prices were less variable. Though all three have had demand elasticities that were as low as or lower (oats and wheat) than corn, their prices were not as variable. Thus, their price elasticities of demand could not be used as an explanation for the difference. The demand for both oats and hay is responsive to changes in livestock production, because their primary use is as feed for animals. The demand for wheat is responsive to changing foreign supply conditions, because a large portion of wheat's demand is due to the existence of export markets. Also, the total production of wheat appears to be more susceptible to climatic conditions than does the supply of oats and hay (Schultz, 1938, pp. 363-371). However, this does not mean that the production of either oats or hay was perfectly stable.

The combination of these factors led to random fluctuations in the prices of oats, wheat, and hay during the late nineteenth century that were not nearly as high as those of corn nor as low as that of cotton. The most reasonable explanation for the differences in price varia-bility may be the existence and development of competitive markets for cotton, with fewer outside factors affecting price fluctuations.

Random Yield Variability of Major Crops

This section discusses the estimates of random variability of crop yields for five major crops—corn, cotton, hay, oats, and wheat— during the late nineteenth century. The random variability coefficients for the leading thirteen producing states of each crop are presented in Table 3-5. (The variability of the number of hogs on farms is also included as a point of reference to the discussion on the corn-hog cycle.)

The estimates in Table 3-5, in general, support the contention that there were regional differences in yield fluctuations. Again, with some exceptions, crop variability was highest in newer agricultural areas.

TABLE 3-5

FIVE MAJOR CROPS (AND HOGS), THIRTEEN LEADING PRODUCERS:
RANKING BY YIELD (AND NUMBER ON FARMS)
RANDOM VARIABILITY COEFFICIENTS

COTTON: Ranking by Yield
Random Variability Coefficients

CORN: Ranking by Yield
Random Variability Coefficients

State	Random Variability. Coefficients (Percent)	State	Random Variability Coefficients (Percent)
Louisiana	24.00	Kansas	31.07
Oklahoma	21.73	Oklahoma	29.93
Texas	20.78	Nebraska	25.98
Florida	18.94	Texas	21.07
Arkansas	17.54	Illinois	19.61
Mississippi	17.09	Missouri	19.20
Tennessee	16.80	Indiana	16.47
Alabama	15.53	Iowa	15.96
South Carolina	15.01	Kentucky	13.69
Georgia	12.98	Georgia	13.01
North Carolina	12.84	North Carolina	12.53
		Ohio	12.17
		Tennessee	11.60

OATS: Ranking by Yield
Random Variability Coefficients

WHEAT: Ranking by Yield
Random Variability Coefficients

State	Random Variability Coefficients (Percent)	State	Random Variability Coefficients (Percent)
Missouri	20.78	North Dakota	27.35
Nebraska	19.95	South Dakota	25.14
Texas	17.19	Indiana	23.47
Kansas	16.78	Ohio	22.82
Illinois	16.43	Kansas	22.30
Indiana	15.21	Illinois	21.64
Iowa	14.99	Missouri	17.26
Minnesota	13.50	Michigan	16.34
Ohio	12.98	California	16.01
New York	12.87	Minnesota	15.88
Pennsylvania	12.78	Pennsylvania	14.30
Michigan	11.06	Iowa	14.08
Wisconsin	10.82	Nebraska	10.68

TABLE 3-5
(Continued)

HAY: Ranking by Yield Random Variability Coefficients		HOGS: Ranking by Number on Farms Random Variability Coefficients	
State	Random Variability Coefficients (Percent)	State	Random Variability Coefficients (Percent)
Kansas	20.09	Nebraska	10.52
Missouri	15.69	Arkansas	5.71
Illinois	15.21	Kansas	5.23
Wisconsin	13.71	Wisconsin	4.83
Virginia	12.19	Texas	4.06
Ohio	11.99	Indiana	3.99
Iowa	11.98	Illinois	3.46
Minnesota	11.87	Iowa	3.05
Indiana	11.65	Kentucky	3.00
New York	10.72	Missouri	2.86
Michigan	10.30	Ohio	2.01
California	7.88	Tennessee	1.73
Pennsylvania	7.07	Georgia	1.48

Source: See the introduction to Appendix A. For any qualifications to the individual estimates, see the notes to the corresponding tables in Appendix A.

The most variable producing states for corn were in the West North Central and West South Central regions. Corn yields tended to vary the least in the East South Central and South Atlantic states. The variability of yields in East North Central states, with the exception of Illinois, was low for corn crops.

Did those same regional differences show up for the other major grain crops? In general, the answer is yes. The variability of oat yields was the highest in West North Central states and the lowest in East North Central and North Atlantic states. Hay yields had about the same level of variability in all of the North Central states, with the exception of Kansas which was higher than the others. The North Atlantic states, which produced hay, and California ranked at the bottom of the list. The variability of wheat yields did not give a picture of regional differences as consistent as the other crops. Some West North Central states' (North Dakota, South Dakota, and Kansas) variability was quite high; whereas, other states from the same region (Minnesota, Iowa, and Nebraska) had very low random yield variability.

The variability of cotton yields was as expected. The most variable producers of cotton were from the West South Central area, with the exception of Florida (a very marginal producer); and the states with the lowest level of variability were all from the South Atlantic region. The East South Central states ranked in the middle of the others listed in Table 3-5. Besides the previously given reasons for most regional differences, the increased use of fertilizer in Georgia, North Carolina, and South Carolina may help to explain their low levels of yield fluctuations (Shannon, 1945, pp. 115; Wright and Kunreuther, 1975, p. 527) and account for some of the regional pattern in the South.

The relative yield variabilities did show some differences between the major crops, falling into two groups. The most variable group was made up of wheat, corn, and cotton. Oats and hay had the lowest variability coefficients. There was much difference between the two groups; though not much difference existed in the variability between the crops within each group. Mean values for the random variability coefficients associated with each crop were calculated from Table 3-5 as follows: wheat, 19.0 percent; corn, 18.7 percent; cotton, 17.6 percent; oats, 14.0 percent; and hay, 12.3 percent.

The level of random variability of yields per acre (as mentioned previously) is a function of uncertain climatic conditions, differing soil and terrain, insect and fire problems, and rapidly changing technologies. Can these unpredictable factors help to explain why the yields of some crops varied more than the yields of other crops? Even though the consequences of these random occurrences were felt by all crops, they obviously were not felt in the same fashion. It may be quite difficult to explain the different levels of variability among the individual crops; however, there appears to be a possible explanation for the sizeable difference between the two groups taken as a whole.

An interesting observation can be made about the random variability coefficients of oats and hay yields, which were much lower than the three other crops. Even though oats and hay obviously were susceptible to climatic and soil conditions, there was no explicit mention of the problems of growing these crops in Schultz (1938). However, a substantial amount of material was written on the problems of growing wheat, corn, and cotton.[11]

Not only is a successful wheat crop dependent upon the yearly amount of rainfall, but the particular seasonal distribution of rainfall is important for the wheat crop to be a success. In fact, climatic conditions are of the utmost concern to wheat growers, because the crop is susceptible to a variety of wheat diseases, if not grown under the right conditions (Schultz, 1938, pp. 363-366). Also, the wrong amount of rainfall or snow covering can be ruinous to yields.

Geographic conditions are more important to corn growers than to wheat producers, because corn depends more upon soil and terrain types than does wheat. In other words, wheat crops can be grown on a wide range of soils; whereas, corn is more limited. Corn yields also may be quite variable, because the crop is so easily damaged in the last two weeks of the growing period (Schultz, 1938, pp. 237-240).

Variations in cotton yields, due to both widely varying soil conditions within the main part of the growing region and differing climatic conditions on the boundaries of the region, may be part of the explanation for high levels of random fluctuations. Because the cotton crop is rather limited by physical conditions to the South and the crop is particularly susceptible to climatic conditions, the uncertainty and great extremes in the weather in this region appear to be reasonable factors for explaining cotton's relatively high yield variability (Schultz, 1938, pp. 285-288).

All in all, the evidence seems to point to the conclusion that some crops (particularly hay) are hardier than others, and some crops (particularly wheat and corn) are more fragile. Considering the large amount of natural uncertainties in the main wheat and corn regions—droughts, floods, prairie fires, blizzards, and locusts, it appears reasonable to expect these crops to have higher random variability co-efficients. Thus, the evidence is close to what one would expect.

Random Income Variability of Major Crops

Farmers ultimately are interested in the net returns per acre and the stability of these returns. They are interested in whether or not returns per acre fluctuate very much about the current level. Because it was impossible to obtain accurate cost data for all forty-eight states for the crops used in this study, gross income variability per acre was used as the measure of fluctuations associated with crop returns--even though returns per acre obviously were dependent not only upon price and yield, but were determined by the interaction of cost, price, and yield.

How accurately does gross income variance measure net income variance? The answer depends upon the nature of costs over time. If the costs of growing a particular crop do not contain any random fluctuations--if they are stable or change predictably, gross income variance will be the same as net income variance. However, this would not be true of the relationship between gross income random variability coefficients and the random variability coefficients of net income. Those two would differ. In fact, because the random variability coefficient is simply the standard deviation as a percentage of the mean, it will be true that our measure of gross income relative variability always will understate net income relative variability as long as costs are stable (or relatively stable). This reasoning follows from the definition of our variability coefficient.[12] If costs contain random fluctuations, then gross income relative variability is not an accurate measure of net income relative variability. This is true because now the variance and the mean of net returns per acre are different from those same parameters of gross returns per acre, because each observation on gross income per acre may have a different cost per acre subtracted from it to calculate net income per acre.

Did these factors cause serious problems? Not too many problems, for several reasons. First, as long as costs were reasonably free from random fluctuations, estimates always were biased downward. Second, as previously discussed, the variate difference method provided an overestimate of the random variance of gross income because of the existence of the product of error and nonerror terms in its final estimate. Third, there was evidence that the random fluctuations in input prices were not nearly as great as the unpredictable fluctuations of prices and yields (Parker, 1972, pp. 402-404; Bowman and Keehn, 1974, p. 603). Thus, the random variability of crop prices and yields might overshadow such movements in cost, making gross income relative variability a reliable approximation (though still lower) of net income relative variability. Last, if it is reasonable to assume that whatever random fluctuations there were in costs, some crops were affected the same--essentially the same inputs were used for these crops--then unpredictable movements in cost would not affect the farmer's choice between crops and cropping systems.

Table 3-6 presents the gross income per acre random variability coefficients associated with the five major crops for the leading producing states of the late nineteenth century. Because gross income per acre is the product of price and yield per acre, the random variability of gross income is a function of the random variabilities of both yield and price. The previous discussion indicated that the year-to-year correlation between prices and yields is important in determining the level of gross income

TABLE 3-6

FIVE MAJOR CROPS, THIRTEEN LEADING PRODUCERS: RANKING
BY GROSS INCOME RANDOM VARIABILITY
COEFFICIENTS

COTTON: Ranking by Gross Income
Random Variability Coefficients

CORN: Ranking by Gross Income
Random Variability Coefficients

State	Random Variability Coefficients (Percent)	State	Random Variability Coefficients (Percent)
Florida	29.90	Missouri	15.97
Oklahoma	21.08	Illinois	15.11
South Carolina	20.66	Oklahoma	14.91
Arkansas	17.94	Nebraska	14.80
Texas	17.66	Kansas	13.15
Louisiana	17.13	Indiana	13.07
North Carolina	16.98	Georgia	12.17
Tennessee	14.66	Ohio	11.44
Missouri	13.47	North Carolina	10.48
Mississippi	12.76	Texas	10.14
Alabama	11.73	Iowa	9.55
Georgia	11.50	Tennessee	8.53
		Kentucky	8.50

OATS: Ranking by Gross Income
Random Variability Coefficients

WHEAT: Ranking by Gross Income
Random Variability Coefficients

State	Random Variability Coefficients (Percent)	State	Random Variability Coefficients (Percent)
Nebraska	17.86	North Dakota	34.90
Minnesota	17.04	Minnesota	28.62
Illinois	16.22	South Dakota	22.77
Texas	15.03	Illinois	20.50
Kansas	15.01	Indiana	20.06
Wisconsin	14.97	Iowa	19.93
Iowa	14.14	Nebraska	18.38
Michigan	14.03	California	17.14
Missouri	13.39	Ohio	16.58
New York	12.73	Missouri	15.84
Pennsylvania	12.29	Pennsylvania	15.54
Indiana	12.04	Kansas	15.32
Ohio	10.35	Michigan	13.96

TABLE 3-6
(Continued)

HAY: Ranking by Gross Income
Random Variability Coefficients

State	Random Variability Coefficients
	(Percent)
Wisconsin	23.24
Iowa	19.10
Virginia	15.09
California	13.69
Michigan	12.91
Missouri	11.17
Illinois	11.11
Minnesota	9.85
Indiana	8.60
Ohio	8.11
Kansas	7.06
New York	6.56
Pennsylvania	4.78

Source: See the introduction to Appendix A. For any qualifications to
individual estimates, see the notes to the corresponding
tables in Appendix A.

variability. Therefore, discussion of these estimates must include information on the year-to-year relationship between changes in prices and
yields associated with the five crops.

Table 3-6 shows regional patterns similar to those previously
observed for price and yield variabilities. Crops grown in the West and
East North Central regions, in that order, ranked one-two in terms of
gross income random variability. North Atlantic states, which were the
leading producers of several major crops, tended to have the lowest level of
income variability. For crops grown in the South, the random variability
coefficients of gross income per acre usually ranked one-two-three for
West South Central, East South Central, and South Atlantic states, respectively. Southern crops, overall, generally ranked in the middle or lower
level of income variability. Nonetheless, there were exceptions, which
can be seen in Table 3-6.

Were there differences in the level of gross income random
fluctuations between the major crops? The answer is yes, as the following
mean values of the random variability coefficients, which were calculated
for the leading producers of the major crops, shows: wheat, 20.0 percent;
cotton, 17.1 percent; oats, 14.2 percent; corn, 12.1 percent; and hay,
11.7 percent.

In the previous discussion, it was noted that wheat yields had the
highest amount of random variability and that wheat crops were in the
middle range of price variability. Yet, as far as returns per acre were
concerned, wheat was by far the most variable. Part of this high variability can be explained by the level of yield random variability, which
showed wheat to be the riskiest crop (as far as yields were concerned).
Nonetheless, that was not the whole explanation. The year-to-year

correlation between changes in prices and yields helps to explain why wheat had the highest random variability coefficients associated with gross income per acre. Because wheat had an insignificant amount of price-yield correlation, yield variability became the most important factor contributing to gross income variability. In other words, income variability of wheat depended solely upon the levels of yield and price variability because prices and yields did not move consistently in the same or opposite directions.[13]

Income variability for cotton producers (taken as a group) ranked second to wheat. The variability associated with yields was moderately high (Table 3-5). Because of the moderate amount of negative year-to-year correlation between prices and yields of cotton, yield variability was the most important factor contributing to gross income variability--that is, the influence of yield variability on income variability was much stronger than the dampening effect of the negative correlation between prices and yields.

The grouping of crops by states masked some important differences among cotton producers. Even though the variability of yields was a more accurate indicator of the level of income variability associated with cotton, it did not give the most accurate indication of the ranking of cotton income variability among states. In fact, the relative ranking among states for income variability was similar to the relative state ranking for price variability associated with cotton. Thus, a state's level of income and yield variability were similar; whereas, that state's level of income variability relative to other states was similar to its level of price variability relative to other cotton-producing states.

Oats and hay both had relatively low levels of yield variability and moderate levels of price variability. Both crops also showed a considerable amount of negative year-to-year correlation between prices and yields (H. Moore, 1923, pp. 18-23). The variability of gross income per acre for these crops was relatively low. Thus, the combination of low yield variability and significant price-yield correlation appeared to explain the estimated levels of income variability for oats and hay. Even though the rankings by yield and gross income variability (Tables 3-5 and 3-6) were somewhat similar for these crops, in some instances, the individual state's yield (and, of course, price) variabilities were poor indicators of gross income variabilities because they ignored significant price-yield correlations.

Corn had the highest level of price variability by far and next to the highest level of yield variability. Thus, as far as price and yield variabilities were concerned, corn was the riskiest late nineteenth century crop. However, an individual farmer ultimately is interested in the variability of a crop's returns per acre. In this respect, how risky was corn? Table 3-6 makes the answer obvious. Gross income random variability coefficients for corn in the major producing states were quite low.

A state's individual price and yield variabilities were poor indicators of gross income variability for corn because they ignored significant and quite high amounts of price-yield correlation. In fact, corn had the highest (negative) estimated correlation coefficient for year-to-year changes in prices and yields (H. Moore, 1923, pp. 18-23). If high prices tend to be associated with low yields and vice versa, gross income variability is reduced. Therefore, even though corn had high individual price and yield variabilities, it tended to have relatively stable returns per acre, because of the year-to-year relationship between prices and yields.

Summary

This chapter presented the estimates of random variability of prices, yields, and incomes associated with several crops and livestock for the late nineteenth century. It was noted that the variability estimates associated with crop series were, in general, a lot greater than those associated with livestock series. Regional patterns were found in the magnitudes of the price, yield, and income random variability estimates. In general, variability associated with each series was greatest in the newer agricultural regions of the country.

An analysis of the variability estimates associated with five major crops grown during the late nineteenth century showed that price and yield variability were not necessarily accurate indicators of gross income variability. For wheat, high levels of both price and yield variability were important factors contributing to high levels of gross income variability. However, this was not true for other crops.

High levels of both price and yield variability for corn were not accurate indicators of gross income variability for that crop. In fact, corn was shown to have relatively low gross income variability due to a significant amount of negative price-yield correlation. Finally, yield variability was usually an accurate indicator of gross income variability for cotton, hay, and oats--though not always.

FOOTNOTES

[1]The following arguments can be drawn from any introductory statistics text, for example, see Yamane (1967) or Yule and Kendall (1965).

[2]Inspection of the computer output shows how far the finite differencing was carried before the systematic element of each time series was eliminated. Thus, the nature of the trend for each time series can be determined to some degree (Yule and Kendall, 1965, pp. 632-633).

[3]Table 3-1 is a cross-sectional summary of representative agricultural states from each region. All estimates are included in Appendix A along with an introduction to the estimates and an explanation of footnotes pertaining to each estimate.

[4]Estimates for all states are included in Appendix A.

[5]See Appendix A for all estimates.

[6]The estimates for North and South Dakota are included in Appendix A.

[7]Estimates for all states of these regions are included in Appendix A.

[8]See Appendix A for any estimates not included in Table 3-6.

[9]The discussion on elasticities was taken from Schultz (1938, pp. 235-462).

[10]For an excellent discussion of the corn-hog cycle, see Shannon (1945, pp. 165-169).

[11]For two elaborate discussions, see Schultz (1938) and Shannon (1945).

[12]We are simply defining the difference between gross income and net income as costs. Because variance is measured as the sum of the squared deviations of an observation from the mean of the sample, if a constant (stable cost) is subtracted from each observation on gross income per acre, then each observation and the mean are smaller by the same amount; thus, the squared deviations are identical. However, the variability coefficient of net income is larger, because it is measured as the ratio of the standard deviation to the mean (which is now smaller).

[13]Henry Moore (1923, pp. 18-23) has calculated correlation coefficients for the year-to-year price-yield relationship during this time period for the five crops as follows: corn, $r=-.78$; oats, $r=-.68$; hay, $r=-.67$; cotton, $r=-.45$; and wheat, $r=-.23$. The correlation coefficient for wheat was statistically insignificant. All others were significant at the 95 percent level of confidence. For tests of the significance of correlation coefficients, see Rao and Miller (1971, pp. 156-158).

CHAPTER IV

CROP DIVERSIFICATION AND RISK IN THE COTTON SOUTH

Introduction

Agrarian history of the American South is a much discussed topic among economic historians. This is not at all surprising, given that the Southern economy remained predominantly agrarian until the twentieth century, and one need look no further than the literature on the topic of cotton and slavery to realize the amount of interest in Southern agriculture.

Recently, scholars have concentrated their efforts on the workings of the cotton economy of the postbellum South. There has been much interest in the topics of cotton "overproduction" and price responsiveness of cotton producers (Ransom and Sutch, 1972; DeCanio, 1973; Brown and Reynolds, 1973; Ransom and Sutch, 1975; Cooley and DeCanio, 1977). The question of productivity and economic retardation in the late nineteenth century South also has been addressed vigorously (Aldrich, 1973; DeCanio, 1974; Wright, 1974). Another topic in which economic historians have shown interest is the influence of risk on the behavior of farmers in the postbellum South (Reid, 1973; Higgs, 1973, Higgs, 1974; Wright and Kunreuther, 1975; McGuire and Higgs, 1977; Wright and Kunreuther, 1977). It is this last topic which is the central concern of this chapter.

Wright and Kunreuther (1975, 1977) recently proposed a particular form of decision-making behavior on the part of Southern farmers to explain what they consider "a major change in the composition of regional output . . . (which) cannot be explained by normal changes in economic incentives" (1975, p. 528). They believe, to put it simply, that there was a "major" shift into cotton and away from foodstuffs (corn) between 1860 and 1880. Their research implied that there was an "over-production" of cotton in the face of stagnating demand by 1880 (1975, p. 526).

A variant of "safety-first" models used by agricultural economists was proposed by Wright and Kunreuther (1975) as an explanation for the change in the cotton/corn output ratio between 1860 and 1880. Their model postulated that Southern farmers had a lexicographical ordering of preferences. In other words, individuals had a discontinuous ordering of multiple goals ranked according to their level of importance. Using this model, Wright and Kunreuther concluded that "institutional and historical developments combined to produce a class of 'gambler' farmers with little to protect by avoiding the risks of the market" (1975, p. 528). That is, they argued that "the rise of tenancy and associated systems of credit" and "the drastic fall in the average farm size" between 1860 and 1880 forced Southern farmers to change their behavior from that of risk averse individuals in 1860 to that of "gamblers" in 1880 (1975, p. 540). Thus, farmers "gambled" by putting increasingly more acreage into cotton.

McGuire and Higgs argued that "the central question that Wright and Kunreuther frame is, in historical perspective, of dubious significance" (1977, p. 168). But that was not the key to their interpretation of the cotton economy of the South. They argued that the empirical evidence provided by Wright and Kunreuther simply was insufficient to determine whether or not postbellum Southern farmers were "gambling" on cotton. Also, they seemed to be reluctant to accept the behavioral assumptions implied by the model advanced by Wright and Kunreuther. McGuire and Higgs suggested that a model based on the straightforward maximization of expected utility might better explain the data (p. 171). They also proposed the use of different statistical measures to determine whether or not postbellum Southern farmers were "gambling" on cotton in the 1880's and whether or not cotton was a riskier crop than corn.

The central concerns of this chapter are as follows: to provide appropriate statistical measures of the variability of prices, yields, and income of cotton and corn crops in the late nineteenth century South; to provide an answer to the question of whether cotton or corn was a riskier crop; to propose the use of a more conventional model to explain the particular crop mix between cotton and corn in the South; and to provide empirical evidence to answer the question of whether Southern farmers were risk averse or whether the particular portfolio of crops chosen by post-bellum farmers implied that they were risk preferers—that is, gamblers.

Before the objectives of this chapter are addressed, it is necessary to know whether or not postbellum Southern farmers were aware of the risks in agriculture. Were farmers aware of long-run trends and cycles? Could they differentiate between systematic and random fluctuations in prices, yields, and income? If so, how responsive were they to the random fluctuations?

If Southern farmers had little information about (or no awareness of) random fluctuations, it would not serve much purpose to study the level of variability in postbellum agriculture. If individuals did not know that they were facing risky choices, their response to such situations could not be measured. If farmers had little or no knowledge about long-run trends and cycles in prices and yields, it would not be relevant to differentiate between systematic and random components of the various agricultural series.

Does the information needed to answer these questions exist? The answer is yes. Postbellum Southern farmers often expressed their concerns

about the risks inherent in agriculture (Carstensen, 1974; McGuire and
Higgs, 1977). Scholars have written about the effects of risk on late
nineteenth century farmers for a long time.[1] And recent studies have
tested directly the relationship between transitory (nonpermanent) changes
in prices and the price-responsiveness of cotton producers of the postbellum South (DeCanio, 1973; Cooley and DeCanio, 1977).

Stephen DeCanio concluded a few years back that "prices did
fluctuate significantly, but adjustments . . . were fairly rapid in most
Southern states. No evidence of alleged traditionalism, or backwardness,
or of merchants' insistence on cotton distorting southern farmers' crop
allocation decisions, can be found in the estimates of cotton supply
functions" (1973, p. 632). More recently, Cooley and DeCanio (1977), using
a varying-parameter estimation method, tested a rational expectations
hypothesis for late nineteenth century agriculture and reached conclusions
having a direct bearing upon the question of farmers' awareness of
uncertainties in agriculture. Their conclusions follow:

> It is obvious, of course, that nineteenth century farmers did
> not mathematically decompose the time series of relative crop
> prices into permanent and transitory components. They are likely
> to be aware of whether relative price changes tended to persist
> or fluctuate randomly. A perspicacious farmer must surely have
> been aware of more of the information contained in the relative
> crop price history than merely the value of the previous year's
> price . . . . But while the precise causes of permanent or
> transitory . . . cotton relative price fluctuations may not have
> been known to the farmers . . . the history of the price fluctua-
> tions was known to them, and our results indicate that a
> substantial portion of the farmers acted on this knowledge . . .
> farmers were neither unresponsive to price changes nor insensitive
> to the history of fluctuations . . . . Our results suggest
> that . . . the farmers displayed a remarkable degree of sophistica-
> tion in their evaluation of . . . [and] . . . perception of and
> adaptation to risks . . . (1977, pp. 15-16; emphases in original).

From this evidence, we believe that postbellum Southern farmers
were aware of long-run trends and cycles, that they could differentiate

between systematic and random fluctuations, that there was an awareness
of risks, and that Southern farmers chose a particular crop-mix by taking
into account the uncertainties associated with that specific portfolio
of crops. Thus, there is definite evidence of the relevance of the
objectives of this chapter.

Measures of Risk for the South

Random variability measures associated with corn and cotton crops
for the Southern states will be presented in this section. Both absolute
and relative random variability estimates of prices, yields, and income, as
well as mean values of each time series, are included in Tables 4-1 through
4-3. All variability estimates for corn were based on forty-four observations from 1866 through 1909, unless otherwise noted. The variability
estimates associated with cotton yields were based on forty-four observations from 1866 through 1909; whereas, the estimates of the variability
of prices and income per acre for cotton were based on forty-one observations from 1869 through 1909, unless otherwise noted.[2]

The variability measures will estimate only that portion of a time
series which is unpredictable from the individual farmer's standpoint.
That is, as previously noted, measures of risk and uncertainty do not
include systematic fluctuations of a series. The variate difference method
will be used to detrend and remove cycles from each time series, because
it is assumed that a farmer has some knowledge about long-run trends and
cycles in prices, technologies, and business conditions. For comparisons
of random variability between series, our measure of _relative_ variability
will be used. It previously was defined as:

$$\text{Random Variability Coefficient} = \frac{\sqrt{\text{random variance}}}{\text{mean}} \times 100$$

where the variance is computed by the variate difference method and measures only random fluctuations of a series.[3]

Price Variability of Cotton and Corn

It seems reasonable to assume that price variability strongly influences an individual farmer's planting decisions. Thus, an investigation into the relative levels of random fluctuations of corn and cotton prices appears useful in determining Souther farmers' behavior under uncertainty. Even though empirical estimates of random fluctuations are not identical to traditional concepts of risk and uncertainty, at least, they are objective measures of the amount of random variability faced by late nineteenth century farmers.

Relative price variabilities of cotton and corn for the cotton states are presented in Table 4-1. The price of corn was more variable by far than the price of cotton. In fact, the random variability coefficients of corn, which were our measures of relative fluctuations, were higher in ten out of the eleven cotton states included in Table 4-1. The only exception was Florida, which was, of course, a quite marginal producer of cotton. From this evidence, it might be concluded that corn was a riskier crop than cotton, at least as far as crop prices are concerned. However, it still might be important to know how cotton price variability ranks relative to other major alternative crops of the region.

Did cotton prices rank high or low in terms of relative variability compared to all major crops? In the cotton states of the South, most crops had higher random variability coefficients than did cotton.[4] Except for the state of Florida, all cotton price random variability coefficients ranked in the lower half of price variability of crops within each state.

TABLE 4-1

CORN AND COTTON PRICE VARIABILITY, SOUTHERN COTTON PRODUCERS:
RANKING BY PRICE RANDOM VARIABILITY COEFFICIENTS
OF CORN*

State	Random Variability Coefficients		Standard Deviations		Means	
	Corn	Cotton	Corn	Cotton	Corn	Cotton
Oklahoma	34.27	17.78	.142	.014	.42	.080
Texas	32.93	10.45	.198	.010	.60	.097
Tennessee	25.58	9.49	.121	.009	.47	.098
Arkansas	22.20	10.04	.127	.010	.57	.099
Louisiana	14.67	10.13	.101	.010	.69	.099
Mississippi	14.20	9.51	.097	.010	.68	.100
South Carolina	13.44	11.63	.103	.012	.77	.101
Alabama	12.57	9.99	.085	.010	.68	.100
Georgia	12.04	7.77	.087	.008	.72	.100
North Carolina	11.39	9.63	.071	.010	.62	.101
Florida	9.66	26.23	.080	.030	.83	.115

Source: See the introduction to Appendix A. For any qualifications as to
the number of observations, see the notes to the corresponding
tables in Appendix A.

* Corn is measured in $/bu. and cotton is measured in $/lb.

And in three states (Arkansas, Louisiana, and Texas), relative price variability of cotton was the lowest of all crops for which there were estimates.

Does the relationship between cotton and corn relative price variabilities in the South appear reasonable? Considering the vastly different conditions determining price movements of the two crops, the difference between the estimated cotton and corn price fluctuations appears quite reasonable and as expected. One has to look no further than the vicious corn-hog cycle and consequent cobweb supply (to the market) and price pattern over time, as described earlier, to expect extreme corn price fluctuations. This factor, combined with a somewhat orderly and quite competitive market for cotton with fewer outside random influences, explains the difference in corn and cotton price variabilities.

Even though farmers ultimately are interested in the variability of crop returns (and not simply price variability) and even more so in the variability of returns from particular crop portfolios, some preliminary observations about the cotton-corn crop-mix of postbellum Southern farmers might be interesting. Because Southern farmers appeared to be quite price responsive and the costs of adjustment to price changes obviously were not zero (Cooley and DeCanio, 1077), it seems reasonable to conclude that some farmers might have been dissuaded from planting corn by the extreme price variability of corn. This would show up in the aggregate as a smaller corn acreage than one might expect from looking only at the returns to corn production. Even if farmers are aware of, adapted well to, and were not insensitive to price fluctuations, it cannot be concluded that individual farmers would want to bear the burdens of these adjustments. Nonetheless, price variability alone cannot explain the particular crop-mix of postbellum Southern farmers. Their actual acreage choice between crops depended on the relationship between the expected returns and the variability of returns of the crop portfolio.

Yield Variability of Cotton and Corn

Yields of all crops fluctuate randomly to some degree because of the uncertainties of weather, natural disasters, and changing technologies. It is only these random variabilities, which are not associated with trend, that are of interest to a study of farmers' behavior under uncertainty. Relative yield variabilities of corn and cotton crops for the Southern cotton states are presented in Table 4-2.

For the overall region, cotton yields tended to have higher levels of random variability than did corn yields. In this respect, Wright and Kunreuther (1975) were correct when they asserted that cotton was the riskier crop of the two. However, closer inspection of Table 4-2 shows that corn yields had the higher random variability coefficient within four states (Georgia, Oklahoma, South Carolina, and Texas). Also, in one other state (North Carolina) no important difference existed between corn and cotton relative yield variabilities. Thus, it cannot be concluded from the evidence on yield variabilities that for all individual states the choice of cotton over corn was necessarily the choice of a riskier crop.

How did cotton yield variabilities rank relative to the random yield fluctuations of other major alternative crops? For most of the eleven states cotton yields ranked in the top half of the yield variabilities associated with all major crops, however, there were four states (Georgia, Oklahoma, South Carolina, and Texas) for which the random yield variability

TABLE 4-2

CORN AND COTTON YIELD VARIABILITY, SOUTHERN COTTON PRODUCERS:
RANKING BY YIELD RANDOM VARIABILITY COEFFICIENTS
FOR CORN*

State	Random Variability Coefficients		Standard Deviations		Means	
	Corn	Cotton	Corn	Cotton	Corn	Cotton
Oklahoma	29.93	21.71	6.671	44.618	24.77	205.56
Texas	21.07	20.78	4.255	40.857	20.19	196.66
South Carolina	18.36	15.01	1.883	26.512	10.26	176.66
Mississippi	14.58	17.09	2.111	31.455	14.48	184.00
Arkansas	14.46	17.54	2.709	37.805	18.73	215.59
Georgia	13.01	12.98	1.364	20.168	10.48	155.34
North Carolina	12.53	12.84	1.590	24.072	12.69	187.50
Alabama	12.12	15.53	1.552	22.455	12.80	144.55
Louisiana	11.63	24.00	1.752	53.114	15.06	221.27
Tennessee	11.60	16.80	2.527	31.856	21.79	189.59
Florida	7.87	18.94	.756	21.748	9.60	114.84

Source: See the introduction to Appendix A. For any qualifications as to
the number of observations, see the notes to the corresponding
tables in Appendix A.

*Corn is measured in bu./acre and cotton is measured in lb./acre.

coefficients of cotton ranked approximately in the middle. In only two states (Florida and Louisiana) was cotton the most variable of all the crops.[5] Therefore, the choice of cotton over all major crops for some states was not necessarily the choice of the riskiest crop with respect to yield per acre.

As expected, both cotton and corn yields had a fair amount of random fluctuations which appeared to be caused, in general, by the great extremes and consequent uncertainty in the weather conditions of the South (Schultz, 1938, p. 287). Yet, because there did not appear to be great differences in yield variabilities between the two crops for many of the states, and because farmers might have responded to yield variabilities by trying to change their techniques of production so as to stabilize yields, no preliminary observations about the particular crop-mix of the South will be made.[6]

Gross Income Variability of Cotton and Corn

Gross income per acre is simply the product of price and yield per acre, and fluctuations in income are, therefore, a function of price and yield variability. Also, because a relationship usually exists between the year-to-year changes in prices and yields, the yearly price-yield correlation may have an important impact on the level of income variability.[7]

In a study of farmers' behavior under uncertainty, only the unpredictable portion of total income variability is desired. Hence, the variate difference method is used to calculate that portion of income variability which from the individual farmer's standpoint is random.[8] Our random variability coefficients for cotton and corn returns are presented in Table 4-3.

TABLE 4-3

CORN AND COTTON GROSS INCOME VARIABILITY, SOUTHERN COTTON PRODUCERS:
RANKING BY GROSS INCOME RANDOM VARIABILITY COEFFICIENTS
FOR CORN*

State	Random Variability Coefficients		Standard Deviations		Means	
	Corn	Cotton	Corn	Cotton	Corn	Cotton
Oklahoma	14.91	21.08	1.418	3.384	9.51	16.05
Mississippi	13.28	12.76	1.279	2.383	9.63	18.67
Arkansas	13.01	17.94	1.365	3.760	10.49	20.95
South Carolina	12.96	20.66	1.004	3.673	7.74	17.77
Alabama	12.30	11.73	1.052	1.698	8.56	14.48
Georgia	12.17	11.50	.903	1.786	7.42	15.52
Louisiana	12.16	17.13	1.240	3.833	10.19	22.38
North Carolina	10.48	16.98	.882	3.179	7.84	18.72
Florida	10.17	29.90	.819	3.787	8.05	12.97
Texas	10.14	17.66	1.190	3.299	11.73	18.69
Tennessee	8.53	14.66	.870	2.746	10.20	18.74

Source: See the introduction to Appendix A. For any qualifications as to the number of observations, see the notes to the corresponding tables in Appendix A.

*Corn and cotton are both measured in $/acre.

When comparing two crops to determine which is riskier, a farmer
ultimately will be interested in the relative variability of the crop returns. The random variability coefficients presented in Table 4-3 show
cotton to be a riskier crop than corn in the postbellum South. The average
value of cotton gross income random variability coefficients for the eleven
cotton states was 17.54 percent; whereas, the same average for corn was
11.83 percent. In this respect, Wright and Kunreuther (1975) were accurate in their assertion that cotton was "risky" in comparison to corn. However, this was not true for every state. In three states (Alabama, Georgia,
and Mississippi), corn had higher levels of income random variability than
cotton. Also, when cotton income is compared to the income variability
associated with all major alternative crops, an interesting situation is noted.[9]

The tables in the appendix show that in three states (Oklahoma, North
Carolina, and South Carolina) cotton ranked in the top half of the income
variability ranking of all crops. Yet it ranked closer to the middle of
the ranking than to the top for these states. In all other states, except
Florida, cotton ranked in the lower half of the income variability ranking
of all crops. Surprisingly, for two states (Alabama and Georgia) only
corn, among all other crops, had lower income variability than cotton.
And, only in Florida (a marginal cotton producer) was income from cotton
the most variable of returns among all crops. Thus, relative to alternative crops grown in the region, cotton was not one of the riskiest with
respect to income per acre.

Why was the random variability associated with corn income, in
general, quite low in comparison to its relatively high levels of price
and yield variabilities? Considering that the overall price variability
of cotton was a lot less than that of corn, why was the income variability

of cotton, in general, higher than that of corn? The answers to these
questions can be explained, in part, by considering the relationship
between the year-to-year changes in prices and yields for each of the
two crops. As mentioned previously, a high degree of negative correlation existed between the year-to-year movements of the prices and yields
of corn during the late nineteenth century. Only a moderate level of
negative price-yield correlation existed for cotton during the same
period.[10] Thus, the high levels of price and yield variability of corn
were not accurate indicators of the level of income variability, because
of the high degree of negative correlation between the year-to-year movements of price and yields of corn. However, because the price variability
of cotton was relatively low and only a moderate level of negative price-
yield correlation existed, the yield variability of cotton appeared to be
the most important determinant of cotton income variability and was an
accurate indicator of the level of income variability.

In this section, concerning the income variability of cotton and
corn, there was the hint that Wright and Kunreuther (1975) might not have
been completely accurate when they asserted that cotton was riskier than
corn. They were accurate when stating that cotton was riskier than corn
for the overall region. However, that was not an accurate statement on a
state-by-state comparison. And even more important is that the evidence
showed cotton to have been one of the least variable crops among the major
alternative crops of the region. Nonetheless, this evidence does not allow
us to answer the crucial question on crop mix: Were Southern farmers risk
takers? An analysis of the farmers' choice of the crop portfolio would be
needed to answer that question. Hence, regardless of whether cotton was

riskier than corn, it cannot be determined solely on the riskiness of the individual crops that postbellum farmers were risk takers--that is, gamblers.

Portfolio Selection

An analysis of farmers' behavior under uncertainty in the postbellum South can be described as a problem in the selection of the appropriate portfolio of crops whose returns are uncertain. If we are interested in the acreage choice of farmers, a model is needed which takes into account the riskiness of the whole acreage portfolio. Risk measures of the individual crops included in a particular portfolio do not determine alone the relative riskiness of that specific crop-mix. Variability of the whole package of outputs also is dependent upon how the crops in the package fluctuate together. Yet, even the determination of the riskiness of one crop-mix relative to the riskiness of another portfolio does not determine whether farmers are risk averse individuals or whether farmers are risk taking (gambling) individuals. Even if one cotton/corn acreage portfolio is riskier than another, it does not follow that if farmers choose the former they would be risk preferers (gamblers). This comparison would only determine the relative variability of prospects.

The determination of risk averting or risk preferring behavior on the part of postbellum Southern farmers depends on having a known preference ranking. In other words, a preference ordering over mixtures of uncertain prospects must be known. The existence of an order-preserving ranking of possible alternatives implies that a given decision rule exists, which allows farmers to choose among alternative portfolios according to that criterion. Because no conclusion about farmers' behavior under uncertainty

can be made without a preference ordering over portfolios, the next logical step in this study is to formulate the model.

The Model

The model which this study tests has a long history.[11] No one
claims it is the most sophisticated model, nor that it always leads to
unambiguous results when applied to all possible prospects. However, the
mean-variance (E-V) approach, which this study will propose, is a straightforward extension of utility theory. Under the assumptions of an E-V
approach, an individual's preference ordering depends solely upon the mean
and variance of returns—an uncertain prospect can be represented fully by
its mean and variance. The decision rule used by a farmer to choose the
appropriate crop-mix from among virtually unlimited possibilities is to
maximize the utility of income derived from the possible acreage portfolios, where utility depends only on the mean and variance of returns.
Our assumption is that the farmers' preference functions can be described,
approximately at least, in terms of the mean and the variance of returns.

There are several reasons why the preference orderings can be
described in terms of these two parameters. One reason is that individuals maximize expected utility and either the underlying utility
function is approximately quadratic in income or the distributions of
returns involve only the mean and variance. In addition, if expected
utility maximization is not assumed, the mean-variance approach still
can be considered as a reasonable first approximation of behavior.
As mentioned in Chapter II, it may be that individuals make a "statistical mistake" or that "people find it easier to compute" or that

distributions facing individuals exhibit little "skewness" or that "information costs" on higher-order moments are prohibitive. For any of these reasons, the mean and variance of returns can become an accurate description of uncertain portfolios.

Harry Markowitz (1959), in an expanded treatment of his original idea, took an approach that ignored the problem of assuming expected utility maximization with a mean-variance analysis.[12] He simply asserted the existence of a utility function of income U(E,V) where

$$\frac{dU}{dE} > 0 \text{ and } \frac{dU}{dV} < 0 \text{ hold.}$$

Our model is based on this assumption.

Given the assumption that $U(E,V)$ exists, our preference function will be linearized for ease of estimation in the following manner:

$$U(E,V) = E - bV,$$

where the "b" represents the subjective risk coefficient of the farmer. This is a great simplification. Yet it should be approximately true, at least locally. Of course, it does lead to easy estimation.

Now the utility of returns to the farmer is a direct function of the mean and the variance of returns. An extension of this model so that the choice object is the maximization of utility of an acreage portfolio is quite simple for a two-crop case. Thus, the farmer makes his acreage decision between only two crops. The equation for portfolio selection is

$$U(Z) = a\mu_x + (1-a)\,\mu_y - b[a^2\sigma_x^2 + (1-a)^2\,\sigma_y^2 + 2(a-a^2)\,\sigma\ xy],$$

where Z represents the returns per acre from a portfolio of two crops—crop x and crop y. The two crops are treated as stochastic variables, where

$$a \geq 0$$

is the fraction of total acres planted in crop x and 1-a is the fraction of
total acres planted in crop y.

$$\mu_x = E(X)$$

and

$$\mu_y = E(Y)$$

represent the expected returns per acre of crop x and the expected returns
per acre of crop y, respectively. In other words, they are the average
return per acre of each crop. The variance of returns per acre of crop x
is σ_x^2 and the variance of returns per acre for crop y is σ_y^2. Lastly,
the covariance of returns per acre of crops x and y is σ_{xy}. Thus, the
expected value of returns per acre for a two-crop portfolio is

$$E(Z) = a \mu_x + (1-a) \mu_y$$

and the variance of returns per acre for the portfolio is

$$\sigma_z^2 = a^2 \sigma_x^2 + (1-a)^2 \sigma_x^2 + 2(a-a^2) \sigma_{xy},$$

where $E(Z)$ and σ_z^2 are simply the mean and the variance of the combination
of two crops, respectively.

There obviously are limitations to the use of this model. Because
only an approximation to the farmer's utility of income function is
represented, the model will give only an approximation of the farmer's
behavior under uncertainty. The literature is replete with discussions of
the problems inherent in the mean-variance approach.[13] According to some
authors,[14] there are only two creditable assumptions which allow for the
use of mean-variance analysis. These assumptions are the existence of a
quadratic utility function or the existence of a normal probability distribution of returns. This view is held because the authors argue in favor

of using expected utility theory in portfolio analysis. Nonetheless, there are still other authors who argue that under a number of reasonable assumptions the mean-variance approach is a useful empirical model (Freund, 1956; Borch, 1968, pp. 50-61).[15]

Estimation Procedure

In this section of the chapter, an attempt will be made to determine if postbellum Southern farmers were risk averting or risk preferring individuals. The mean-variance model discussed will be used for this determination. Because theoretical models are not always identical to empirical ones, our empirical test must be specified clearly.

For purposes of estimation, our assumptions are straightforward. The linearized preference function is

$$U = E - bV.$$

There are no assumptions about how a farmer weighed the variance of returns. In fact, because risk averting and risk taking behavior are measured by the farmer's subjective risk coefficient, the objective of the estimation procedure is to obtain an estimate of b. It should be noted that the utility function is specified in terms of mean-variance per acre. The assumption here is that farmers were interested in returns per acre of crops grown. This is a standard assumption in the literature (Carter and Dean, 1960, pp. 190-195).

For acreage portfolios consisting of two crops (cotton and corn), the model will be of the following form:

$$U = a\mu_x + (1-a)\mu_y - b[a^2 \sigma_x^2 + (1-a)^2 \sigma_y^2 + 2(a-a^2) \sigma_{xy}],$$

where all variables are defined as previously, x represents corn, and y represents cotton. The farmer is assumed to have maximized U. Thus, the farmer's decision was to choose an "a" (his crop-mix) that would lead to this maximization. It follows that the first order condition for the maximization of U will be

$$\frac{dU}{da} = \mu_x - \mu_y - b[2a\sigma_x^2 - 2(1-a)\sigma_y^2 + (2-4a)\sigma_{xy}] = 0.$$

From the first order condition, the solution of "b" is straightforward:

$$b = \frac{\mu_x - \mu_y}{2[a\sigma_x^2 - (1-a)\sigma_y^2 + (1-2a)\sigma_{xy}]}.$$

The subjective risk coefficient, b, is used to determine whether or not farmers were risk averse. If the estimated coefficient is

$$b > 0,$$

then it can be concluded that the farmers were risk averse individuals, because they tried to avoid variance. However, if the estimated coefficient is

$$b < 0,$$

then one might conclude that the farmers were risk preferers, because in this case they actually had a preference for higher levels of variance at a given level of expected return. Lastly, if the estimated coefficient is not significantly different than

$$b = 0,$$

it would appear that the postbellum Southern farmers were risk neutral individuals, as they placed no subjective weight on the variance of returns and simply maximized expected return.

While the model does provide a criterion to determine risk aversion or risk preference on the part of postbellum farmers—by the sign of the subjective risk coefficient, the model does not provide an objective criterion on which to evaluate the magnitude of each estimate.

This is true because the "b" is an estimate of the farmers' <u>subjective</u> attitudes toward risk. Nevertheless, most of the estimates should be positive because empirical observations lead one to believe that individuals are, in general, risk averse. However, no knowledge exists about how large the estimates "should" be.

The calculation of the risk coefficients was carried out by using annual state data for the years 1869-1909[16] for nine cotton states--North Carolina, South Carolina, Georgia, Florida, Alabama, Mississippi, Arkansas, Louisiana, and Texas.[17] Ideally, individual farm data would have been used; however, those data did not exist. Thus, the use of annual state data became the next best alternative. But, because estimates were based upon state-wide aggregates, inferences about any individual farmer's behavior under uncertainty cannot be drawn from the results. Nonetheless, because Wright and Kunreuther (1975) argued that postbellum Southern farmers as a group were "gamblers," we <u>can</u> address their conclusions. Furthermore, the existence of a certain type of behavior under uncertainty in the aggregate would lead one to believe that that type was predominant among the Southern farmers.

Because, as previously mentioned, farmers are assumed to have some knowledge about economic and technological trends and cycles, the variate difference method was used to calculate the variance of returns associated with cotton and corn and the covariance of returns between cotton and corn. Random variance and random covariance (determined by the variate difference method) were used for the calculation of b. Only the unpredictable portion of total variance of returns was relevant to a farmer's decision under uncertainty.

Estimations of the risk coefficients for the nine states were carried out at eight-year intervals for each state. The purpose of subperiodization was twofold. One reason was that without some type of subperiodization the estimates of the risk coefficients would imply (on the part of farmers) a forty-one year subjective weighting scheme and would be relevant only for a farmer who made an acreage decision in 1910. Because this was unrealistic, a more reasonable alternative was sought. One assumption was that a declining weighting scheme exists with respect to time. That would imply a much shorter interval than forty-one years. Others have argued, with reasonable success, that a declining weighting scheme is appropriate for late nineteenth century agriculture (Fisher and Temin, 1970; DeCanio, 1973; Cooley and DeCanio, 1977).

The eight-year intervals were chosen mainly on statistical grounds for purposes of calculating the random variances. Also, most models that assume the existence of declining weights would place a negligible weight on observations past the eighth year. Nonetheless, experimentation was carried out with respect to the length of intervals to determine if the results would change significantly. No important differences were noted by changing the interval one or two years (plus or minus). Thus, the eight-year interval was chosen.

The other reason for sub-periodization was for use in tests of significance. A distribution of the risk coefficients (b's) was needed for the determination of their statistical significance. The model's inability to provide an objective criterion with which to evaluate the magnitude of the subjective risk coefficient made a formal test necessary. Also, as our coefficients were only estimates, the results would not be meaningful without a formal test.

Extensive experimentation was carried out with different eight-year intervals for the entire period under study. Five eight-year intervals were estimated starting in 1869 and ending in 1908. Two more sets of eight-year intervals were estimated starting in 1870 and 1871.[18] These three sets of eight-year intervals were examined for the existence of significant differences in their estimates of the random variance and covariance terms. Essentially, we were trying to determine if the data were very sensitive to the choice of time intervals. Because the estimated random variances and covariances were similar, for the most part, between their respective overlapping eight-year intervals, our risk coefficients were estimated using the eight-year intervals starting at the beginning of the sample period--1869. Nonetheless, there was no way around the obvious fact that the choice of those particular intervals was arbitrary.

Implicit in this analysis was that subjective probability beliefs were held by the farmers for the likelihood that the returns from an acreage portfolio would assume alternative values and that these beliefs were determined by past experiences. To estimate the subjective risk coefficients (b's), therefore, the expected return of corn (μ_x); the expected return of cotton (μ_y); the variance of corn returns (σ_x^2); the variance of cotton returns (σ_y^2); and the covariance of corn and cotton returns (σ_{xy}) were estimated at eight-year intervals starting in the following years: 1869, 1877, 1885, 1893, and 1901. It was assumed then that the estimated parameters determined the farmers' choice of crop-mix--their "a's"--for the following years: 1877, 1885, 1893, 1901, and 1909. To calculate the acreage proportion in corn--the value of "a," the total harvested acreage in corn and cotton was assumed to equal 100 percent. Harvested corn acreage then was calculated as a proportion of the total acreage. That determined our "a." The proportion of total harvested acreage in cotton equaled "1-a."

The next step was to substitute the estimates into the following equation (which is the same as previously given):

$$b = \frac{\mu_x - \mu_y}{2[a\sigma_x^2 - (1-a)\sigma_y^2 + (1-2a)\ \sigma_{xy}]}.$$

This procedure was repeated for all nine states for the five eight-year intervals. Thus, there were forty-five estimates of "b."

Results

The statewide estimates of the farmers' subjective risk coefficients (b's) for all five eight-year intervals are presented in Table 4-4. On examining Table 4-4, one cannot help but notice the obvious. All forty-five of the estimates were positive. This was a striking result considering the simplicity upon which the estimates were founded. Nonetheless, a formal test of significance was carried out as a means of solidifying the results. However, the test of significance was less than was desired.

Because distributions of the individual state estimates of "b" did not exist, the best that could be done was to test the mean of a set of risk coefficients. In other words, a test of the significance of the difference from zero of an individual b was not possible without a known distribution of that state's subjective risk coefficients. To test for the significance of the difference of the mean of b from zero, we made an assumption of normality and proceeded with a t-test.[19] The null hypothesis to be tested was that

$$H_0 : \mu_b = 0$$

TABLE 4-4

ESTIMATES OF SUBJECTIVE RISK COEFFICIENTS, SOUTHERN COTTON PRODUCERS: RANKING BY STATES

State	b_{1877} Based on 1869–1876	b_{1885} Based on 1877–1884	b_{1893} Based on 1885–1892	b_{1901} Based on 1893–1900	b_{1909} Based on 1901–1908
North Carolina	3.896	1.130	2.529	3.230	2.779
South Carolina	.538	1.139	6.165	1.320	.602
Georgia	.890	1.260	3.816	2.906	1.224
Florida	4.233	.376	.699	1.017	2.581
Alabama	1.826	.765	1.554	1.858	2.666
Mississippi	1.120	1.023	2.995	1.295	6.030
Arkansas	.475	.899	6.426	1.301	8.037
Louisiana	.727	1.023	1.715	1.483	.770
Texas	1.590	.701	6.861	1.425	.316

Source: See Appendix B.

and the alternative hypothesis was that

$$H_A: \mu_b > 0.$$

Thus, the null hypothesis was that the mean of the population of b's was zero. The alternative hypothesis against which the null was tested was that the mean of the population of b's was, in fact, non-zero and positive. Because we had an estimated mean from a sample of b's, our interest was to test whether the sample mean was significantly different from the population mean.

The appropriate t-statistic for this test was

$$t = \frac{\bar{b} - \mu_b}{\dfrac{\sigma_b}{\sqrt{n}}}$$

where $\bar{b}$ was the sample mean, μ_b was the population mean (=0), σ_b was the unbiased estimate of the sample standard deviation, and n was the number of observations in the sample of b's.

Using the estimates of the subjective risk coefficients from Table 4-4, allowed the testing of two hypotheses about the b's. As Wright and Kunreuther had argued that Southern farmers' attitudes toward risk changed over time due to changing institutions, the first hypothesis was that although the b's for a given year (for example, 1877, 1885, 1893, 1901, or 1909) came from a homogeneous population--all nine states made up this population--the estimates over time came from different populations. That is, the test of significance would be on the means of b's of all nine states in a given year--the columns of Table 4-4. That provided five tests of significance. The other hypothesis about the risk coefficients was that although the estimates across states did not come from a homogeneous population--the b's of farmers within a given state were from

the same population, but they were different than the population of b's
from any other state--the subjective risk coefficients were from a
homogenous population over time. The tests of significance, in this
case, were on the means of b's of each state across all eight-year
intervals--the rows of Table 4-4.

Because no a priori information existed that would lead us to
believe which of the two hypotheses was the truth, the necessary tests
were carried out for both hypotheses. The results of these tests are
presented in Tables 4-5 and 4-6.

The results in Table 4-5 show the computed t-statistics to be
quite large in comparison to the critical value of t at the 95 percent
level of confidence. Thus, the difference between $\bar{b}$ (the sample mean)
and μ_b (the population mean which is set equal to zero) is significant,
and the null hypothesis for this test was rejected. That is, under the
assumption that the postbellum Southern states were a homogeneous group
at a given point in time (as far as their behavior under uncertainty was
concerned), we could conclude that evidence existed that farmers in those
states behaved as risk averse individuals. That conclusion held for all
five of the eight-year intervals tested.

Table 4-6 presents the results of t-tests under the other assumption. In this case, it was assumed that the farmers' attitudes toward
risk were homogeneous over time within a given state and that each state
could be viewed as representing different (heterogeneous) populations. For
six out of the nine states, the results were such that the difference
between $\bar{b}$ and μ_b was significant, and the null hypothesis that the population mean of b's was equal to zero was rejected. Again, there exists evidence for these six states, that the farmers were risk averse individuals.

TABLE 4-5

T-STATISTICS FOR THE TEST OF MEANS OF THE SUBJECTIVE RISK COEFFICIENTS
OF ALL STATES IN A GIVEN YEAR: RANKING BY YEAR

Year	Interval Upon Which Estimates Are Based	t-statistics	$\bar{b}$	σ_b	$\sqrt{n}$
1877	1869–1876	3.598*	1.699	1.417	3
1885	1877–1884	10.146*	.924	.273	3
1893	1885–1892	4.716*	3.640	2.316	3
1901	1893–1900	6.789*	1.759	.778	3
1909	1901–1908	3.163*	2.778	2.634	3

Note: $\bar{b}$ is the sample mean, σ_b is the sample standard deviation, and n is the number of observations in the sample.

*Null hypothesis rejected at 95 percent level of confidence.

TABLE 4-6

T-STATISTICS FOR THE TEST OF THE MEAN OF THE SUBJECTIVE RISK
COEFFICIENTS OF A GIVEN STATE ACROSS TIME:
RANKING BY STATE NAME

State	t-statistic	$\bar{b}$	σ_b	$\sqrt{n}$
North Carolina	5.913*	2.713	1.026	2.236
South Carolina	1.836	1.953	2.379	2.236
Georgia	3.542*	2.019	1.275	2.236
Florida	2.472*	1.781	1.611	2.236
Alabama	5.680*	1.734	.683	2.236
Mississippi	2.610*	2.493	2.136	2.236
Arkansas	2.110	3.333	3.533	2.236
Louisiana	5.833*	1.144	.439	2.236
Texas	1.826	2.179	2.669	2.236

Note: $\bar{b}$ is the sample mean, σ_b is the sample standard deviation, and n is the number of observations in the sample.

*Null hypothesis rejected at 95 percent level of confidence.

However, for three states (South Carolina, Arkansas, and Texas), the results were such that the null hypothesis that the mean of the population of b's was equal to zero could not be rejected. Thus, it is _possible_ that Wright and Kunreuther were right when they asserted that Southern farmers' attitudes toward risk might have been changing during the postbellum period. But this is not strong evidence for their case, because when the null hypothesis cannot be rejected, it does not follow that it is accepted. Also, the results indicated that the null hypothesis cannot be rejected in _only_ three states out of nine.

What was shown? Can any substantive conclusions be made? It does appear that cotton was a riskier crop than corn. Yet with this increased variability came an increased expected return.[20] It appears that the Southern farmers were aware of both. The particular crop-mix that they chose does not indicate gambling behavior. In fact, the results of the utility maximizing model used here indicate that postbellum farmers would have had to increase their cotton acreage substantially or would have had to receive a substantially lower expected return from cotton to indicate risk preferring behavior.[21]

All forty-five estimates of risk coefficients were positive. Of the fourteen separate tests of the coefficient means, all but three indicated that the sample means were significantly different than zero and positive. For those three tests, the results indicated simply that the samples were such that _no_ conclusions were warranted. We believe that these results indicate that postbellum Southern farmers behaved quite rationally and were, in fact, risk averse individuals. Furthermore, the large difference in expected returns between cotton and corn was the very factor that made the chosen acreage portfolio a rational one.[22]

In addition, we believe our results are strong enough to show that Wright and Kunreuther (1975) were mistaken in their assertion that postbellum Southern farmers were "gamblers."

FOOTNOTES

[1]For some textbook accounts of the effects and importance of risks and uncertainty on the behavior of nineteenth century farmers in general, see Shannon (1945), North (1966), Parker (1972), and Gray and Peterson (1974).

[2]State yield series for cotton exist for the whole period under study (1866-1909) while the state cotton price series did not start until 1876 and had missing observations in 1877 and 1881. An average United States cotton price was used from 1869 to 1875 and for the years 1877 and 1881 in place of the actual state cotton price for each series; therefore, the income series exists only from 1869 to 1909. No average United States cotton price series exists before 1869. For a detailed discussion on the relationship between U. S. price and state prices, see the introduction to Appendix A. Also, any exceptions to the time periods are in the appendix.

[3]For a detailed discussion of the points in this paragraph, see the previous chapter. Also, for an excellent discussion on the difference between absolute and relative measures of dispersion, see Yamane (1967, pp. 75-77).

[4]Price variability estimates for all major crops are included in Appendix A.

[5]Yield variability measures for all major crops are included in Appendix A.

[6]An individual farmer in a price-taking market has virtually no control over the market price he receives for his crop. Thus, he may respond to extreme price variability by avoiding that crop, particularly when facing market demand which is price inelastic. But the individual farmer does have some control when faced with extreme yield variability; for example, the particular technique of production.

[7]For a detailed discussion of the concepts in this paragraph, see the section in the previous chapter on income variability.

[8]See above note.

[9]Gross income variability estimates for all major crops are in Appendix A.

[10]Henry Moore (1923, pp. 18-23) estimated the coefficient of correlation of price-yield relationship for corn as -.78 and the coefficient of correlation for cotton as -.45, based on 35 observations.

[11]See Chapter II in this study, which discusses the risk and uncertainty literature; particularly the section on portfolio theory.

[12]As discussed in Chapter II, Markowitz took his approach because he apparently considered expected utility theory to be controversial; particularly the consistency conditions.

[13]See Chapter II for a detailed discussion of the problems in E-V analysis.

[14]For an excellent discussion of this point, see Feldstein (1969).

[15]An additional reason to those presented in Chapter II and earlier in this chapter on why the mean and variance of returns may be used to describe the alternative prospects is that Feldstein's (1968) "statistical mistake" and the existence of prohibitive costs of obtaining information on the whole probability distribution might be the same problem. Also, Borch (1968, pp. 60-61; 1969, p. 4) uses examples of "dramatically different" distributions that would prevent the use of E-V analysis; however, the distribution of returns from acreage portfolios faced by Southern farmers were not "so dramatically different" that they would invalidate the Markowitz model. This study simply agrees with the conclusion made by Borch that even with its simplification of real-life decisions, "the Markowitz model may correspond fairly well to actual business practice" (1968, p. 52).

[16]For an explanation of the data, see footnote 2 above and the introduction to Appendix A.

[17]Because the model postulates that the returns from only two crops—cotton and corn—are relevant to the farmers' acreage decision, the states chosen must be essentially two-crop states. However, it is obvious that there are no states that grow only two crops. Thus, states that have an overwhelming proportion of their total agricultural acreage planted in cotton and corn may be viewed as essentially two-crop states. Because data on acreage planted do not exist for the period under study, data on harvested acreage were used. During the postbellum period, all but three of the states chosen—North Carolina, South Carolina, and Georgia—averaged having 90 percent or more of their total harvested acreage in cotton and corn; South Carolina and Georgia averaged having about 85 percent of their total harvested acreage in cotton and corn. Even though North Carolina averaged having only 68 percent of its total harvested acreage in cotton and corn, it could be included in the study, because no other crops were effective alternatives for the farmers making a decision about the cotton-corn mix. Yet the state of Tennessee was excluded, because cotton was simply too minor a crop relative to other major crops grown in the state.

[18]The five eight-year intervals, starting in 1871, obviously, end beyond the period of this study, in 1910. However, as will be mentioned, this set of eight-year intervals was not used.

[19]Even though it may be argued that there was no a priori reason for assuming that the b's come from a normal distribution, particularly in light of the small sample at hand, this assumption should not cause any problems of interpretation. It has been shown that a t-test is not sensitive to the assumption of normality. For the derivation, see Henry Scheffe, The Analysis of Variance (1959, p. 337).

[20]See Appendix B.

[21]For a detailed discussion of these points, see the introduction to Appendix B.

[22]See note 21.

CHAPTER V

AGRARIAN DISCONTENT AS A RESPONSE TO UNCERTAINTY

Background

The economic basis of agrarian discontent in American agriculture during the last third of the nineteenth century is a much discussed topic. In this chapter an attempt will be made to apply our variability estimates to the problem. An hypothesis will be proposed that argues that agrarian discontent was a response to "high" levels of uncertainty in American agriculture.

The traditional view of the farmers' protest movements is that much of the unrest was due to worsening economic conditions in agriculture. For farmers, this view centered around complaints of monopoly pricing by grain elevators, railroads, moneylenders, land speculators, and big business. The farmers perceived these practices as leading to higher prices for goods and resources purchased by farmers. Farmers also complained that the monetary policy of government caused deflation in agricultural prices received without a corresponding decline in prices paid, thus causing a decline in agricultural terms of trade.[1]

In contrast to the traditional view is that held by economic historians conducting recent research on the economic basis of agrarian unrest. It has been argued that declining agricultural terms of trade could not adequately explain agricultural discontent, because the trend in the terms of trade, in general, was relatively stable or improving (Bowman, 1963; North, 1966; Bowman and Keehn, 1974). There has been evidence against the existence of exploitation in output and labor markets which could have caused unrest in the South (DeCanio, 1973; 1974). Railroad rates might help to explain some of the unrest in particular states during certain periods. However, the general trend of the relative price of railroad transportation (which has been shown to have been quite stable) did not appear to be an adequate explanation of the discontent as a whole

(Higgs, 1970). The argument that monopoly moneylenders charged excessive

interest rates to farmers does not appear to hold up in the face of recent

evidence (Bogue, 1965; Bowman, 1965). Lastly, it does not appear that

monopoly land speculators exploited farmers (Bogue, 1963).

Other scholars have argued that even though farmers really were

not exploited by markets in the way the traditional view perceives the

situation, these farmers were facing economic distress, because their

understanding of existent economic conditions was poor. It has been

suggested that nineteenth century farmers were bewildered by the new

economic environment and that it was the increasing amount of adjustments

to changing conditions that burdened the farmers (North, 1966; Mayhew,

1972).[2] Even so, others have produced evidence that appears to contradict

Mayhew and North's suggestions. The view central to this opposing

argument is that the farmers were aware of economic reality and responded

properly to their economic environment (Fisher and Temin, 1970; DeCanio,

1973; Cooley and DeCanio, 1977). To quote Cooley and DeCanio on the results

of these research efforts:

> Economic studies of the price-responsiveness of late-nineteenth
> and early twentieth century American agriculture have shown that
> the sector as a whole responded properly to market prices in both
> the choice of crop-mix and the choice of technique. . . . This
> paper goes beyond the previous studies by testing directly a
> rational expectations hypothesis for American agriculture . . . we
> will show that changes in the farmers' price expectations were
> indeed consistent with the theory of rational expectations (1977,
> p. 9; emphasis in original).

Even though Cooley and DeCanio's results implied that some underlying

behavioral changes on the part of the farmers took place during the

Populist period, they still concluded that farmers' behavior was quite

rational.

In light of this evidence, can one find an economic basis for the farmers' discontent? A few of the recent studies on the subject have suggested a possible common basis for the unrest. Bowman and Keehn, in their study of agricultural terms of trade, presented evidence that "shortrun fluctuations, not a secular decline in the terms of trade, may have been a major cause of farmers' discontent during the last decades of the nineteenth century" (1974, p. 594). They further claimed the existence of "a close relationship between fluctuations in the terms of trade and the rise and fall of the major agrarian protest organizations" (p. 603). Their results showed substantial year-to-year variation in the terms of trade for the four states (Illinois, Indiana, Iowa, and Wisconsin) studied. In essence, it appears that Bowman and Keehn felt that uncertainty was, in fact, a major source of farmer unrest. Yet that was not the proposition actually tested. They compared cyclical fluctuations in the terms of trade with the cycles inherent in the protest movement. They simply studied the rise and fall of the terms of trade over short intervals.

Klepper (1973) also found that the cyclical behavior of some economic variables appeared to coincide with the rise and fall of the protest movements. Yet, he went one step further than Bowman and Keehn by directly testing—using a multiple regression analysis—the effect of the variability of gross income on the protest movements. Again, a scholar was suggesting that agrarian discontent might have been a response to "high" levels of uncertainty in agriculture. Yet even Klepper was not actually testing that proposition, because his explanatory variable was not a measure of risk (or uncertainty). It was only a measure of the variability of the raw data. Also, his dependent variable was a protest voting index—which, admittedly, could be a rather poor measure of agrarian discontent. In fact, he admitted that some of his poor results were probably due to the use of voting as a measure of farmer unrest.

One economic historian directly suggested that the protest movements might have been caused, in part, by the amount of uncertainty in agriculture.[3] Robert Higgs asserted:

> Greater sources of unrest were the extreme instability and consequent unpredictability of farm production . . . these random occurrences affected differently the various parts of the supplying area . . . farmers generally understood the reasons for this instability but could neither control nor reduce it (1971, pp. 101-102; emphases added).

Higgs further stated that in the economic profession, scholars "now recognize that certainty itself is for most people an economic good" (1971, p. 102).

If certainty is an economic good for most people, then individuals, in general, would prefer to have more of it rather than less. Yet, if farmers "could neither control nor reduce" the amount of uncertainty facing them, how are they to "purchase" more certainty? Is it possible that farmers of the late nineteenth century felt that government action could help them in this matter? If so, how could they persuade the government to help? Is it possible that protest activity was simply the least-cost method of convincing government to help? We feel that farmers were aware of the extreme risks in agriculture, that few solutions existed in the marketplace, that farmers perceived that the government—by changing the "rules of the game"—could help, and that farmers felt that protest activity was their best means to a solution.

It is in question whether the government really could have created more certain conditions in agriculture; it appears that authors holding the traditional view of farmer discontent believe it could have done so.

Hicks (1961, pp. 87-90 and pp. 92-95) repeatedly claimed that if the government had moved to a flexible greenback monetary system, stable business conditions would have followed. He further argued that this action would have been a partial solution to the problems of agriculture. Holding similar views, Shannon (1945, p. 316) argued that it is a "fact that gold alone was" not a sufficient basis for coinage. He continued by arguing that the farmers were right in their claim that a bimetallic standard, or the free and unlimited coinage of silver alone, would have led to a stable price level. Again, an historian viewed government action as part of the solution to uncertainty in agriculture.

Could it be possible that the extreme instability in farming was really a major source of unrest? Obviously, uncertain conditions in agriculture were not the only source, but unstable prices and business conditions did seem to be a prevalent topic encompassed in contemporary statements. One Illinois farmer, Willard G. Flagg, wrote in 1874 that one of the major causes of farmers' unrest was as follows:

> As a combined result of the opportunities afforded by vicious legislation . . . speculation and a large amount of uncertainty pervades all business. Monopolies founded on the special privileges granted by special legislation are made more permanent . . . . The Farmers' Movement in the Western States means, then, . . . a recurrence of one of those periods not uncommon in history of unusual oppression and distress caused by bad government, and resulting in rebellion on the part of the oppressed; and finally, an effort to reverse the unwise legislation that has, in the guise of corporate and other monopolies, created, fostered, and perpetuated a Shylock aristocracy . . . (emphases added).

In 1875, a railroad man and frequent contributor to magazines on railroad subjects, Charles Francis Adams, offered his opinion on the causes of the Granger movement. Speaking about the railroad system, he concluded:

> The people of the West . . . were . . . to realize that competition between its members was producing results neither such as had been anticipated nor such as were altogether satisfactory. They

found, in a word, that while the result of ordinary competition
was to reduce and to equalize prices, the result of railroad
competition was to produce local inequalities and to <u>arbitrarily</u>
<u>raise and depress prices</u> (emphases added).[4]

The preceding passages express the farmers' frustrations and concerns

over the existence of uncertain business conditions. The remedy expressed

by Willard G. Flagg was to turn to government.

Late nineteenth century farmers also were convinced (though perhaps

quite wrongly) that speculation on their crops and government monetary

policy created extremely uncertain conditions (Buck, 1913; 1920). That the

farmers wanted these situations, at least, corrected is evidenced by many

of the demands they made:

> That the law-making powers take early action on such measures
> as shall effectually prevent the dealing in the future of all
> agricultural products . . . (Demands of the Farmers' Alliance
> of Texas, 1866).

> That we demand the abolition of national banks, and the substitu-
> tion of legal tender treasury notes in lieu of national bank notes,
> issued in sufficient volume to do the business of the country on a
> cash system; <u>regulating</u> the amount needed . . . . That we demand the
> free and unlimited coinage of silver (The St. Louis Demands, 1889;
> emphasis added).

> We demand a national currency, safe, sound, and <u>flexible</u>, issued
> by the general government only . . . (The People's Platform, 1892;
> emphasis added).[5]

James Weaver, in an article written in 1892, also clearly expressed that

the monetary policy of the government, because of its great uncertainty,

was a major source of distress to farmers.[6]

Scholars, for a long time, have stressed the high degree of risk in

agriculture. Some have placed a great deal of emphasis on this fact

(Shannon, 1945; Higgs, 1971; Parker, 1972); and others have mentioned it only

in passing (North, 1966; Carstensen, 1974).[7] However, there has yet to be

a test of the relationship between the uncertain conditions in agriculture

and farmer discontent. This study attempted such a test.

Introduction to the Tests

Because neither of the two hypotheses we formulated had a strong theoretical foundation, the tests used were, obviously, not very rigorous. The tests consisted of some rather simple comparisons of our estimates of price, yield, and income variability with the historical evidence on the location and timing of agrarian discontent. Nonetheless, because the objective of this chapter was to test how well our risk measures could help to clarify an important topic in American history, the results should be interesting.

The first approach, which was rather straightforward, was a cross-sectional analysis. If, during the late nineteenth century, a major source of farmer unrest was the uncertain conditions in agriculture, then our risk measures should show certain state differentials, because, according to the historical literature, there were obvious geographical patterns to the protest movement. Thus, the test was to compare a cross-sectional analysis of the variability measures for the whole period to a cross-sectional analysis of discontent.

The second approach to the problem was a time-series analysis. There was evidence of cycles in both the amount of fluctuations in prices, yields, and income facing the farmers and in the level of protest activity during the late nineteenth century. Research efforts by Bowman and Keehn (1974) and Klepper (1973) suggested a connection between cycles in fluctuations and the rise and fall of various protest movements. Thus, our approach was to periodize the data so as to correspond with the timing of protest activities and then test whether our variability measures over those intervals increased and decreased with the rise and fall, respectively, of protest activities. In essence, we tested for the homogeneity of variance with respect to time.

The Cross-Sectional Analysis

For a variety of reasons, our analysis concerned only New York, Pennsylvania, and all states in the North Central region; one important reason being that evidence on the geographical differences in protest activity appeared best for states in the North.[8] In addition, even though the South and Far West were involved in protest activity, the main strength of the unrest remained in the North (Bowman and Keehn, 1974). There also was evidence that farmers in the South truly faced a different set of problems than farmers in the North. Thus, one should not lump both regions together (Mayhew, 1972). Lastly, because the activity in the South was temporally more limited and much of the agrarian unrest in the North was more widespread over time, a comparison of activity in the North Central region with our variability estimates for the whole period appeared more plausible.[9]

During the period under study (1866-1909), four major reform movements--Granger, Greenback, Alliance, and Populist--attracted discontented farmers. The Alliance movement was purely agrarian in nature; while the Populist and Granger movements were heavily agrarian; and the Greenbackers, partly agrarian in nature. Nevertheless, all four movements were the major political voice of agrarian protest at various times throughout the last half of the nineteenth century.

The Granger movements became strongest after 1867 and rose to their peak about 1874. They then proceeded to decline in importance by 1880 (Bowman and Keehn, 1974, p. 603). Much agreement exists in the literature that the states of Illinois, Iowa, Minnesota, Wisconsin and possibly Nebraska were the most active Grange states during this period

(Buck, 1913; Shannon, 1945; Unger, 1964; Bowman and Keehn, 1974). Although many other Northern states had Granger activities within their farming communities, they did not have much success at passing the so-called Grange legislation which regulated railroads and grain elevators. Thus, these states were not considered as having been very active as protest states.

Greenbackers--unlike the Grange organizations, which were primarily interested in regulation of railroads, grain elevators, other "middlemen," and monopolies--were interested mainly in money issues. They wanted repeal of the National Banking Act and return of control over the money supply to the federal government. They wanted the government to partake in the issue of paper money (greenbacks) for inflationary purposes. They hoped for more stable prices, in general, and for higher agricultural prices, in particular. Rural interest in the greenback issue did not start until after the panic of 1873 and was strongest in the late 1870's, with rapid decline by the early 1880's. States most active in the Greenback movement as a form of agrarian discontent were Illinois, Indiana, Iowa, and Wisconsin (Unger, 1964, p. 384). Agrarian greenbackism also was prevalent in Minnesota and Missouri, although not as much as in the preceding group of states (Unger, 1964, p. 385). Several other states were strongly involved in the Greenback movement. States such as Michigan, Ohio, and Pennsylvania supported Greenbackism. Yet the support in these states came from various labor and business interests in iron and mining areas. Greenbackism in places such as these could not be considered part of agrarian discontent (Unger, 1964, pp. 255-271).

Several Alliances and Leagues were formed in the Midwest during the late 1870's and 1880's. Although not as active as earlier movements in terms of agrarian protests, these new organizations eventually served as a major political voice for discontented farmers (Bowman and Keehn, 1974, p. 604). The Alliance movement was quite strong as a vehicle for agrarian protest by the late 1880's in the states of Iowa, Kansas, Minnesota, Nebraska, North Dakota, and South Dakota (Hicks, 1931, pp. 103-148).[10] Hicks (1932, p. 100) also mentioned that the Alliances were popular in Illinois, Michigan, Missouri, and Wisconsin in the late 1870's and very early 1880's, but they were never strong as a protest movement.

After the elections of 1890, the height of agrarian unrest was reached. The Populist Party was formed in the early 1890's and remained the most vociferous farmer movement until nearly the end of the century. It is fairly clear from the literature on farm discontent that the Populists were most active in the states of Kansas, Nebraska, North Dakota, and South Dakota during the 1890's (Hicks, 1931, pp. 225-265; Shannon, 1945, pp. 317-326). Although other Midwestern states contained some Populist activity, Populist strength, in terms of actual protest, was relatively small (Hicks, 1931, pp. 274-300; Higgs, 1971, p. 89).

From the above information concerning the various farm movements during the latter part of the nineteenth century, can one determine which states were the most active and which states were the least active in terms of agrarian discontent? We believe this can be done. Because the Dakotas were active in both the 1880's and 1890's and our sample period for them begins in 1882, they appear to be two likely candidates for the most active category during the period under study. Also, the states of Kansas and Nebraska were active throughout almost the entire period and thus rank in the most active category. Ranking as the next most active

were Illinois, Iowa, Minnesota, Missouri, and Wisconsin, for at some time
each was quite active and at other times, very inactive, in the agrarian
protest movements. States such as Indiana, Michigan, and Ohio could be
considered, at best, in a marginal category, while New York and Pennsylvania (even though still important producers of some crops) had virtually
no agrarian unrest relative to the other states.

If the above evidence is accepted as indicative of a reasonable
ranking of protest activity among the states, we now are ready to test
for the existence of a relationship between levels of uncertainty and
differential levels of agrarian discontent across states. Our random
variability coefficients calculated by the variate difference method will
be used as the basis for comparison between states. They will indicate
the relative levels of uncertainty facing farmers with respect to prices,
yields, and income. The variability measures associated with the four
most important crops of this region--wheat, corn, oats, and hay--are
included. Those estimates are for the whole period under study (1866-
1909), except for states with shorter recorded histories, these exceptions
being the estimates for Minnesota (1867-1909), North Dakota (1882-1910),
and South Dakota (1882-1909). The price variability estimates associated
with the four crops are presented in Table 5-1, yield variability estimates
in Table 5-2, and income variability estimates in Table 5-3.

An inspection of Table 5-1, which ranks the states according to
their level of price variability, shows some interesting results. The
table is similar to a ranking of the relative level of protest activity
among the states. Kansas and Nebraska appeared to have the highest level
of random price variability when all four crops are considered, while

TABLE 5-1

WHEAT, CORN, OATS, AND HAY; FOURTEEN NORTHERN AND NORTH CENTRAL
STATES: RANKING BY PRICE RANDOM VARIABILITY COEFFICIENTS

State	Wheat Random Variability Coefficient	State	Corn Random Variability Coefficient
North Dakota	24.43	Kansas	49.08
South Dakota	24.09	Nebraska	48.83
Nebraska	20.38	Missouri	39.93
Kansas	19.75	Iowa	33.03
Iowa	18.39	Illinois	30.25
Wisconsin	16.98	South Dakota	26.64
Minnesota	16.27	Indiana	20.21
Missouri	15.79	Ohio	18.16
Illinois	15.03	North Dakota	14.67
Indiana	13.88	Minnesota	11.33
Michigan	13.46	Michigan	10.18
Ohio	13.25	Pennsylvania	9.75
Pennsylvania	12.06	Wisconsin	9.46
New York	11.11	New York	9.24

State	Oats Random Variability Coefficient	State	Tame Hay Random Variability Coefficient
Illinois	28.08	Kansas	27.84
Nebraska	27.54	Nebraska	24.68
Kansas	25.05	Missouri	24.59
South Dakota	24.58	Wisconsin	24.42
Iowa	21.08	Illinois	19.86
Missouri	19.48	North Dakota	17.59
Wisconsin	16.88	Indiana	15.82
Michigan	16.67	Michigan	15.49
Minnesota	16.19	Ohio	13.62
New York	15.38	Minnesota	13.11
Indiana	14.54	New York	10.23
Pennsylvania	14.40	Iowa	10.13
Ohio	14.99	South Dakota	6.74
North Dakota	11.72	Pennsylvania	6.73

Source: See the introduction to Appendix A. For any qualifications to
individual estimates, see the notes to the corresponding tables
in Appendix A.

states such as Michigan, Ohio, New York, and Pennsylvania had the lowest
levels with respect to the four crops. Of the remaining states, Illinois,
Iowa, Missouri, and South Dakota had the highest levels of price variability.
In addition, if oats variability is ignored for North Dakota and hay variability for South Dakota (both crops are very minor in these states), then
both of the Dakotas ranked quite high in terms of price variability.

In fact, if one ranked each state's level of price variability
only with respect to the more important crops in the state, then price
variability was even more closely related to the level of protest activity.
For example, the most important crop in North and South Dakota was wheat,
and those two states had the highest levels of wheat price variability
of all the states and, of course, were quite active in the farm protest
movements. States with virtually no protest activity, such as Ohio, New
York, and Pennsylvania, had very low levels of price variability in their
major crops—wheat, oats, and hay. Kansas and Nebraska, major producers
of corn and wheat, had the highest levels of corn price variability of all
states, and next to the Dakotas, the highest levels of wheat price variability. Both states also were quite active in the protest movements. In
general, these results were encouraging.

What kind of results did we get from the estimates of yield variability? Were they as encouraging as the price variability results? The
yield estimates did support the hypothesis that agrarian discontent was a
response to uncertainty in agriculture. However, the results were not
quite as convincing as the previously mentioned price variability estimates.

The random yield variability measures, as presented in Table 5-2,
indicated that the Dakotas, Kansas, and Nebraska had by far the highest

TABLE 5-2

WHEAT, CORN, OATS, AND HAY; FOURTEEN NORTHERN AND NORTH CENTRAL
STATES: RANKING BY YIELD RANDOM VARIABILITY COEFFICIENTS

State	Wheat Random Variability Coefficient	State	Corn Random Variability Coefficient
North Dakota	27.35	Kansas	31.07
South Dakota	25.14	South Dakota	26.54
Indiana	23.47	Nebraska	25.98
Ohio	22.82	North Dakota	22.07
Kansas	22.30	Illinois	19.61
Illinois	21.64	Missouri	19.20
Missouri	17.26	Indiana	16.47
New York	16.55	Iowa	15.96
Michigan	16.34	Minnesota	13.55
Minnesota	15.88	Wisconsin	13.50
Pennsylvania	14.30	Ohio	12.17
Iowa	14.08	Michigan	11.24
Wisconsin	11.97	Pennsylvania	10.24
Nebraska	10.68	New York	9.08

State	Oats Random Variability Coefficient	State	Tame Hay Random Variability Coefficient
South Dakota	25.95	Kansas	20.09
North Dakota	22.94	Nebraska	15.76
Missouri	20.78	Missouri	15.69
Nebraska	19.95	Illinois	15.21
Kansas	16.78	North Dakota	13.98
Illinois	16.43	Wisconsin	13.71
Indiana	15.21	South Dakota	12.79
Iowa	14.99	Ohio	11.99
Minnesota	13.50	Iowa	11.98
Ohio	12.98	Minnesota	11.87
New York	12.87	Indiana	11.65
Pennsylvania	12.78	New York	10.72
Michigan	11.06	Michigan	10.30
Wisconsin	10.82	Pennsylvania	7.07

Source: See the introduction to Appendix A. For any qualifications to
individual estimates, see the notes to the corresponding tables
in Appendix A.

levels of fluctuations with respect to all four crops, while New York, Pennsylvania, and Michigan had the lowest. Of the remaining states, Illinois and Missouri appeared to have the highest overall levels of yield uncertainty. When each state was ranked according to the level of yield variability with respect to its most important crops, conclusive results were obtained for only some of the preceding states. Those states highest in variability--North and South Dakota, Kansas, and Nebraska--were also the most active in protest movements, while the states lowest in yield variability--New York, Pennsylvania, and Michigan--were the least active in terms of agrarian discontent. However, the results were inconclusive concerning other states.

The states of Ohio and Indiana, major producers of wheat and corn during the late nineteenth century, had very high levels of wheat yield variability and very little protest activity. The state of Iowa, which was at times heavily involved in agrarian discontent, had moderate to low levels of yield variability for its major crops. Thus, it appears that the relationship between uncertainty in agriculture and farmer unrest may be less certain with yield than with price variability estimates.

Ultimately, farmers are concerned with the net income variability of alternative crops. Income variability, in turn, is determined by the relationship between yield, price, and cost. However, the impossibility of obtaining cost data associated with the major crops across states over a forty-four year historical period necessitated the use of gross income data in computing crop income variabilities. As mentioned earlier, though, we are of the opinion that the relative nature of income variability between crops (or across states) was not seriously biased.

The gross income variabilities of the four crops are presented in Table 5-3. Income variability estimates, like the yield estimates, did not match the level of protest activity among the states as well as did the price variability coefficients. North Dakota, Nebraska, Wisconsin, and probably Illinois had the highest levels of income variability over all crops while there was little doubt that Michigan, Ohio, Pennsylvania, and New York consistently ranked near the bottom of variability with respect to the four crops. The latter states clearly would not be considered part of the protest movements; however, only North Dakota and Nebraska of the former states would be considered as heavily involved in agrarian discontent.

If comparison of the income variability measures and protest activity was limited only to the most important crops within a state, the results tended to improve. In those states most active in agrarian discontent--North Dakota, South Dakota, Kansas, and Nebraska, the variability estimates of their most important crops often ranked in the top half among all states, while the income estimates of those states least active in agrarian unrest--Indiana, Ohio, New York, and Pennsylvania--usually ranked in the bottom half for their most important crops. One obvious exception to this comparison was the income variability of wheat in Kansas. Kansas, a major protest state, had next to the lowest level of wheat income variability of all states. Thus, the relationship between income uncertainty and protest activity was positive for many states, but negative for others.

While the comparison of price, yield, and income variability measures with the level of activity of states in farm protest did not give <u>completely</u> conclusive results, our comparison did strongly support the

TABLE 5-3

WHEAT, CORN, OATS, AND HAY; FOURTEEN NORTHERN AND NORTH CENTRAL
STATES: RANKING BY GROSS INCOME RANDOM
VARIABILITY COEFFICIENTS

State	Wheat Random Variability Coefficient	State	Corn Random Variability Coefficient
North Dakota	34.90	Wisconsin	17.62
Minnesota	28.62	North Dakota	16.50
South Dakota	22.77	Missouri	15.97
Illinois	20.50	Illinois	15.11
Indiana	20.06	Nebraska	14.91
Iowa	19.93	Kansas	14.80
Wisconsin	18.45	Indiana	13.07
Nebraska	18.38	South Dakota	12.54
New York	16.89	Ohio	11.44
Ohio	16.58	Pennsylvania	10.94
Missouri	15.84	Minnesota	10.43
Pennsylvania	15.54	New York	10.07
Kansas	15.32	Michigan	9.58
Michigan	13.96	Iowa	9.55

State	Oats Random Variability Coefficient	State	Tame Hay Random Variability Coefficient
North Dakota	27.65	Wisconsin	23.24
Nebraska	17.86	Iowa	19.10
Minnesota	17.04	North Dakota	17.76
Illinois	16.22	Nebraska	14.10
Kansas	15.01	Michigan	12.91
Wisconsin	14.97	Missouri	11.17
Iowa	14.14	Illinois	11.11
Michigan	14.03	Minnesota	9.85
Missouri	13.39	South Dakota	9.20
South Dakota	13.05	Indiana	8.60
New York	12.73	Ohio	8.11
Pennsylvania	12.29	Kansas	7.06
Indiana	12.04	New York	6.56
Ohio	10.35	Pennsylvania	4.78

Source: See the introduction to Appendix A. For any qualifications to
individual estimates, see the notes to the corresponding tables
in Appendix A.

contention that agricultural uncertianty was a major source of farmer discontent during the late nineteenth century. In addition, there did exist
several reasons why the results might have been less than conclusive. The
sources of farm discontent came from many factors only one of which might
have been the uncertainty inherent in agriculture. Thus, a simple comparison between variability measures and protest activity could not be expected
to yield conclusive results. Also, because our measures of variability included at least ten years which were relatively free of protest activity
(the estimates were based on a sample period that ended in 1909), those
random variability estimates obviously were not the "best" measures of the
uncertainty facing the farmers involved in agrarian protest movements.

To correct for some of the preceding problems in our explanation
of protest activity across states, the data were sub-periodized, and the
variability measures were computed for the shorter intervals. Using the
historical literature on protest activity as the basis for the choice of
periods, we estimated variability coefficients for the following eight-
year intervals:[11] (1) Granger movement, from 1867 through 1874; (2) Greenback movement, from 1874 through 1881; (3) Alliance movement, from 1883
through 1890; and (4) Populist movement, from 1890 through 1897. Price,
yield, and income random variability coefficients associated with wheat,
corn, and oats grown in the fourteen states also were computed. These
estimates are presented in Tables C1 through C12 in Appendix C.

Did these additional estimates support the hypothesis that agrarian
unrest was a response to the extreme instability in agriculture? It is
interesting that these variability measures did show the existence of large
differences across states during each sample period for most of the series
being estimated. However, the estimates did not always match perfectly

with the level of unrest within a state. Nonetheless, the eight-year estimates did provide additional support to the hypothesis. Although the strength of this support may be a matter of opinion, we believe it is quite strong.

The estimates presented in Tables C1 through C3 were computed for the period of Granger unrest (1867-1874). Variability estimates for the states most active in the Granger movement—Illinois, Wisconsin, Iowa, and Minnesota—were not always the highest. However, the price variabilities for these states ranked for the most part in the top half among all states. Their yield variabilities ranged from nearly the highest to nearly the lowest. In terms of income variabilities, though, these states ranked very high. In fact, with the exception of income variability associated with oats in Illinois, their computed income variability coefficients ranked approximately in the top third among all states. It is interesting to note that the two states which overall had the highest levels of variability during the Granger period—Nebraska and Kansas—had the "largest ratio of granges to the total farming population" at the peak of the movement (Shannon, 1945; p. 310).

The variability coefficients presented in Tables C4 through C6 were computed for the Greenback period (1874-1881). Because evidence on the strength of individual state's participation in the Greenback movement as a form of agrarian discontent was not abundant, the significance of the computations for this sample period may not be perfectly clear. Nevertheless, from an extensive study of the subject, one can conclude that farmers from Illinois, Iowa, and Indiana were heavily involved in Greenbackism and that farmers in Kansas, Minnesota, and Wisconsin also were quite involved (Unger, 1964, pp. 345-407). An analysis of the variability coefficients representing the three crops in these six states showed more than 75 percent of the total number ranking in the top half of the estimates for all states. Of these variability measures, most (more than 75 percent) were in the highest third among all states' estimates.

Random variability coefficients computed for the Alliance period (1883-1890) are presented in Tables C7 through C9. From the historical literature on the subject, enough information can be derived to determine that the strength of the Alliance movement was in the states of Kansas, Nebraska, Minnesota, and the Dakotas (Hicks, 1931, pp. 128-152; Hofstadter, 1955, pp. 94-109). The price, yield, and income variability measures of the three crops for these states, in general, tended to be at the higher end of the variability scales. In fact, of the total estimates for the five states, nearly 70 percent were in the top half of the variability range. Their price variabilities usually ranked near the highest in relative variability, while their yield estimates ranged from the highest to near the bottom of the variability scale. However, the income variability estimates for the three crops (excepting corn in South Dakota) grown in Kansas, Nebraska, and the Dakotas were at the highest end of the variability scale. Crops grown in Minnesota ranked in the middle range of relative variability of income. Yet, if one of the five states could have been considered less active in the Alliance movement, it would have been that state.

Did we receive any additional information from an investigation of the Populist period? We did have fairly clear evidence that Kansas, Nebraska, the Dakotas, and possibly Minnesota were heavily involved in Populist discontent (Hicks, 1931, pp. 255-263, 328-329), and an analysis of the price, yield, and income variabilities contained in Tables C10 through C12 supported the hypothesis that agrarian unrest was a response to the extreme instability in agriculture. For the Populist period (1890-1897), the computed variability estimates of the three crops ranked near the highest overall in Kansas, Nebraska, and the Dakotas. Minnesota ranked relatively high only in terms of price variability, while its yield and income variabilities were relatively low. However, the analysis did have some problems. The yield and income variabilities for the three crops were among the highest in both Illinois and Indiana. Also, the price variabilities in Illinois were among the highest. Because neither state could be considered as being heavily involved in the Populist movement, it appears that extreme instability and agrarian discontent were not <u>always</u> closely related.

The results of sub-periodization suggested that uncertainty in late nineteenth century agriculture might have played an important part in agrarian discontent. Those states most active in the protest movements of a given sample period--for example, Kansas, Nebraska, and the Dakotas during the Populist period--usually had the highest levels of relative variability among all states. Even though the sub-period results for these states have not been discussed in detail, Tables C1 through C12 clearly show that in the areas of relatively little or no protest--Michigan, Ohio, Pennsylvania, and New York--the variability measures for wheat, corn, and oats were consistently at the bottom end of each variability scale. Nonetheless, some problems of interpretation do arise, because the relationship between protest activity and variability levels in the other states is less clear. In other words, because the relationship between agrarian unrest and the level of uncertainty facing farmers, though often positive, was sometimes negative, our results would have to be considered less than <u>completely</u> conclusive.

The Time-Series Analysis

This section examines the possibility that changes in price, yield, and income uncertainty over time accounted for differences in the levels of discontent with respect to time. It has been well established that the level of protest activity within states rose and fell during the last third of the nineteenth century (Bowman and Keehn, 1974). If the large amounts of instability and consequent uncertainty inherent in agriculture actually played a causal role in agrarian unrest, then one would expect our measures of random variability for a state to change through time. However, before a time-series test could be formulated, a concise framework upon which to base the test had to be established. This framework was founded on the description of the activity and timing of selected states' agrarian unrest presented below:[12]

Illinois. Evidence exists of this state's involvement in both the Granger (1867-1874) and Greenback (1874-1881) movements. Although farmers in Illinois obviously were members of Alliances (1883-1890), that was not a period of major unrest in the state. Also, little protest came from Illinois during the Populist period (1890-1897).

Wisconsin. This state appears to have followed the same pattern as Illinois--strong protest within the first two periods with relatively little during the latter two.

Iowa. The state of Iowa had a pattern of agrarian discontent quite similar to the preceding two states. Even though farmers in Iowa were part of the Alliance movement, it could be argued that the level of political activity was much less than in the earlier periods. Little protest was, no doubt, the case during the Populist period.

Minnesota. Farm groups in Minnesota were nearly always at a high level of political activity. Although agrarian unrest may have been somewhat less during the later periods than the earlier ones, there do not appear to be substantial differences as there were in the preceding three states.

Indiana. The evidence suggests that Indiana was active in farmer unrest only during the Greenback period. Little protest came from this state during the periods from 1867-1874, 1883-1890, and 1890-1897.

Nebraska. Although active earlier, Nebraska had its major periods of farm protest during the Alliance and Populist movements. Thus, on a relative basis, little protest came from farm groups in Nebraska during 1867-1874 and 1874-1881.

Kansas. Farmers in Kansas followed a pattern of protest similar to Nebraska farmers. Relatively little protest during the two earlier periods and quite strong unrest during the later two periods.

North Dakota. North Dakota had the same pattern as Kansas and Nebraska.

South Dakota. South Dakota had the same pattern as the three preceding states.

Alabama. Even though some agrarian agitation appeared during the Greenback period among Southern states, the thrust of farmer discontent was at its height during the Alliance and Populist periods. Thus, farm groups in Alabama were quite active politically in 1883-1890 and 1890-1897, while little agrarian protest took place in the South during the periods 1867-1874 and 1874-1881.

Georgia, Louisiana, Mississippi, North Carolina, South Carolina, Arkansas, and Texas. Farm discontent among these states followed a pattern similar to that in Alabama. Vociferous protest came, in general, from the South during both the Alliance and Populist movements, while relatively little agrarian protest came out of the South during the Granger and Greenback movements.

If these descriptions of agrarian protest activity in the selected states are reasonably accurate, then the results of our tests can be expected to conform, in general, to the following framework:

1. For the states of Illinois, Wisconsin, and Iowa, the variability measures of the several series to be estimated should be higher in both the Granger and Greenback periods than during the time of both the Alliance and Populist movements. Also, the random variability measures computed for the former two periods should be similar to each other, while these same measures computed for the latter two periods should be similar to each other.

2. For the state of Minnesota, the variability measures of the several series to be estimated should be similar for all four periods, because agrarian discontent was, in general, at a relatively high level during each period.

3. Because farm groups in Indiana were involved in protest movements only during the Greenback period, there should be lower variability measures for the state during all other periods. Thus, variability measures of the several agricultural series for the Granger, Alliance, and Populist movements should be at similar levels, while those same measures should be significantly higher during the Greenback movement.

4. For Nebraska, Kansas, the Dakotas, and all selected states in
   the South, the level of variability of their several agricultural series should be higher during the Alliance and Populist
   periods (and similar between the two periods) than during the
   Granger and Greenback movements (which should have similar
   levels of variability between them).

Because this part of the study was an attempt to determine whether
uncertainty in agriculture played a causal role in the timing of various
protest movements, a determination of the homogeneity (with respect to time)
of the unpredictable variability facing farmers during the late nineteenth
century was needed. Thus, several agricultural series were sub-periodized
and computations were made of random variances over those shorter intervals.
A test of the significance of the difference between random variances computed from a given series then was made to determine homogeneity with respect to time.

The justification for a test of the homogeneity of random variances,
not the random variability coefficients which had been used as our measures
of relative riskiness, was founded on a restrictive assumption. Because
the units of measurement of the series being compared were the same, no
problems were created; however, it has been assumed implicitly that the
means of each sub-period series being compared were the same. While this
assumption was restrictive, it did allow for the use of a straightforward
test of the equality of variances. Additionally, as a practical consideration, there seldom existed significant differences between the means of the
sub-periods from the same series.[13]

The procedure used for determining the homogeneity of the random
variances with respect to time was the F-test.[14] The null hypothesis to

be tested was that

$$H_0: \sigma_1^2 = \sigma_2^2$$

and the alternative hypothesis was that

$$H_A: \sigma_1^2 > \sigma_2^2,$$

where σ_1^2 represented the random variance of the series under study during a particular time period (for example, the Populist period) and σ_2^2 represented the random variance of the series under study during a different time period (for example, the Granger period). Thus, our null hypothesis was that the population random variances of a particular agricultural series during two different time periods were equal. The alternative hypothesis that the null was tested against was that the population random variance of a particular agricultural series during one of the time periods was greater than the population random variance of the same series during the other time period.

The appropriate F-statistic for this test was

$$F = \frac{\hat{\sigma}_1^2}{\hat{\sigma}_2^2},$$

where $\hat{\sigma}_1^2$ and $\hat{\sigma}_2^2$ were unbiased estimates of σ_1^2 and σ_2^2. Also, when constructing the F-statistic, the larger variance always went in the numerator. Thus, the test statistic used here was simply the ratio of our estimate of the random variance of a series during one period to our estimate of the random variance of the same series during a different period. Finally, our F-test was based on a 95 percent level of confidence.

Results

Price, yield, and gross income random variances associated with four crops—wheat, corn, oats, and cotton—were computed for the seventeen selected states during the Granger (1867–1874), Greenback (1874–1881), Alliance (1883–1890), and Populist (1890–1897) periods.[15] These estimates are presented in Tables C13 through C24 in Appendix C.

The F-tests were carried out according to the historical framework discussed earlier. The procedure followed was to interpret the historical framework as an hypothesis about the equality or inequality of specific random variances and then to use the F-test to determine whether statistically significant differences existed between the estimated random variances. If the computed F-statistic exceeded the critical F (3.79), the null hypothesis that the two variances were equal was rejected. This tended to support the hypothesis that heterogeneity of the random variances with respect to time existed, thus lending support to the hypothesis that uncertainty in agriculture played a causal role in the timing of agrarian unrest. For example, the framework implied that for Illinois, the random variances of the several series should not be significantly different between the Granger and Greenback periods, nor should they be significantly different between the Alliance and Populist periods. However, significant differences in the random variances would be expected between the Greenback and Alliance periods, with the variances from the latter period being significantly less than those from the former period. This procedure was carried out for the several series of each state. We did not expect that every series would conform to the implied historical hypothesis, yet we did expect many to conform. However, the results were rather discouraging.

The results of the F-tests, which were in agreement with the historical description of protest activity among the selected states and which lend support to a causal role of uncertainty in the timing of the protests, are listed below (the number in parentheses is the value of the computed F-statistic):

Random Variances of Wheat, Corn, Oats, and Cotton Prices. None.

Random Variances of Wheat Yields. For Iowa, the estimate fell significantly from the Greenback to the Alliance period (10.48).

Random Variances of Corn, Oats, and Cotton Yields. None.

Random Variances of Wheat Income. For Indiana, the estimate fell significantly from the Greenback to the Alliance period (34.79). For Kansas, the estimate rose significantly from the Greenback to the Alliance period (4.29).

Random Variances of Corn and Cotton Income. None.

Random Variances of Oat Income. For Illinois, the estimates fell significantly from the Greenback period to the Alliance period (5.08).

If we accept, as Bowman and Keehn (1974) suggested, that the Alliance period actually contained substantially less political activity by farm groups than the other periods, then the results listed below also could be considered as supportive of the causal role of uncertainty in the timing of agrarian unrest (the number in parentheses is the value of the computed F-statistic):

Random Variances of Cotton Prices. For North Carolina, Georgia, Alabama, Mississippi, Arkansas, Louisiana, and Texas, the estimate rose significantly from the Alliance to the Populist period (15.00, 6.50, 7.50, 15.00, 17.00, 21.00, and 15.00, respectively).

Random Variances of Wheat Yields. For North Dakota, the estimate rose significantly from the Alliance to the Populist period (3.85).

Random Variances of Corn Yields. For South Dakota, the estimate rose significantly from the Alliance to the Populist period (4.77).

Random Variances of Oat Yields. For Kansas, Nebraska, and Alabama, the estimate rose significantly from the Alliance to the Populist period (5.47, 6.20, and 4.23, respectively).

Random Variances of Cotton Yields. For Alabama, Mississippi, Arkansas, and Texas, the estimate rose significantly from the Alliance to the Populist period (6.69, 4.76, 5.08, and 4.20, respectively).

Random Variances of Wheat Income. For South Carolina, the estimates rose significantly, from the Alliance period to the Populist period (5.04).

Random Variances of Corn Income. For South Dakota and Alabama, the estimate rose significantly from the Alliance period to the Populist period (4.18 and 14.06, respectively).

Random Variances of Oat Income. For Georgia, the estimate rose significantly from the Alliance period to the Populist period (6.48).

Random Variances of Cotton Income. For Alabama and Arkansas, the estimate rose significantly from the Alliance period to the Populist period (4.95 and 5.25, respectively).

Taken as a whole, these time series results have to be considered as less than satisfactory. It does not appear that our variability measures explained the timing of agrarian discontent very well. Not only did very few of the random variances move in the same direction as our historical descriptions, but many of them moved in the opposite direction. In fact, only four variances moved in the same direction as our original expectations, while ninety-five moved in the opposite direction of our original expectations. Even when we used our revised expectation about the random variances during the Alliance movement, only twenty-six variances moved in the correct direction while seventy-three variances moved in the opposite direction.[16]

Why did the time-series analysis yield such poor results? Was it because of faulty estimating procedures, or was it because of a faulty historical description of agrarian discontent? The choice of an eight-year interval was arbitrary, and one could argue that this choice might have adversely affected the results. Yet we could not have chosen a much shorter interval and had meaningful random variance estimates. If much longer intervals had been chosen, there would have been overlap between the several protest movements whose temporal patterns we were attempting to explain. Nevertheless, experimentation was carried out on the use of different intervals, and this did not lead to significant changes in the results.[17]

The historical description of farmer discontent during the late nineteenth century given here is reasonable. While it probably is true that some scholars would not agree, we believe most scholars familiar with the subject would agree. However, the description may be considered reasonable only in a general sense and may not necessarily be an an accurate description of the exact time and place of each movement. Thus, the poor results may be accounted for, in part, by this problem. Yet we do not think so.

Another important factor in accounting for the poor results may be that an historical event as complex as the agrarian uprisings of the last third of the nineteenth century cannot be explained easily as a response to any single factor. Farmers clearly were not very pleased with the extreme instability inherent in agriculture. But an attempt to determine a causal role of one factor—uncertainty—without accounting for the causal roles played by other factors, may not be a satisfactory approach. Klepper (1973), in his extensive study of the subject, argued essentially that an economic basis of farmer discontent was much more complex than economic historians had previously thought.

The most important factor contributing to the nature of the time-series results was that the annual state observations, with which the variance ratios were computed, were such that no statistically reliable conclusions could be made. With only seven degrees of freedom, one needs a very large absolute difference between the variances to show that a "significant" difference exists. From our results, we cannot reject the null hypothesis that the variances were equal over time for the greater part of the tests. Yet that does not imply that we accept it. All that is known is that the differences might be considered due to chance. In conclusion, our random variances were such that they were not completely satisfactory for a time-series analysis.

FOOTNOTES

[1]For an excellent discussion of the traditional view, see Hicks (1931). Also, see Buck (1913, 1920), Shannon (1945), or Carstensen (1974).

[2]The conclusion that North and Mayhew really were arguing that farmers' perceptions and responses to reality were faulty was originally stated by Cooley and DeCanio (1977). Cooley and DeCanio, however, cited North's 1974 second edition of Growth and Welfare as the reference for his view. Yet North can be "fairly" accused of holding the position of the farmers' bewilderment only by reference to his 1966 first edition of Growth and Welfare, because his position changed in the 1974 edition.

[3]Although this position was stated originally by Higgs (1971), the view also is held by Douglass North as the following quote by him makes clear: "What was fundamentally at stake in the farmer's discontent was, first of all, that he found himself competing in a world market in which fluctuations in prices created great uncertainty" (1974, p. 134; emphases added).

[4]Both articles, as cited in Carstensen (1974, pp. 53, 57-58).

[5]All of the above, as cited in Carstensen (1974, pp. 73-74, 77-78, 89-93).

[6]This article is also cited in Carstensen (1974, pp. 80-89).

[7]In his second edition, North (1974) definitely emphasized that uncertainty was a major source of distress. Yet there was no test of the proposition; see footnote 3.

[8]Most textbook accounts of agrarian discontent clearly agree on which states in the North were most active and which states had relatively little protest activity. See Gray and Peterson (1974), Higgs (1971), or North (1974). Also, the traditional accounts of farmer unrest are in agreement about the location of discontent. See Buck (1913, 1920), Hicks (1931), or Shannon (1945).

[9]Of course, states in the North were not heavily involved in the agrarian movement all of the time. Yet, as will be argued later, some of the states could be described as protest states; whereas, others could not. According to the literature, states in the South were heavily involved in protest activity, in general, only in the late 1880's and 1890's. See Hicks (1931), Shannon (1945), or Bowman and Keehn (1974).

[10]Also, on the strength of Alliances in different states, see Shannon (1945, pp. 311-326).

[11]The use of an eight-year sample period also was based on statistical considerations. Much shorter sample periods would have led to meaningless estimates of the random variance.

[12]The selection of the states presented here was determined on the basis of evidence in the historical literature discussed previously in this chapter. Only those states whose protest activity could be determined (and timing of their activity known) were chosen. The basic references for this selection were Buck (1913), Hicks (1931), Shannon (1945), Hofstadter (1955), Saloutos (1960), Unger (1964), and Carstensen (1974).

[13]An additional constraint on the general applicability of the test proposed—the F-test—was that it was based on an assumption of normality which if violated can be serious; see Yamane (1967, pp. 642-654).

[14]The use of an F-test for testing the equality of variances with respect to time necessitated the exact specification of the relationships between the several sub-period variances. For this reason, we had to be rigid in our historical descriptions of the time and place of the various protest movements. See Yamane (1967, pp. 642-654) for a detailed description of the F-test.

[15]The choice of time periods was discussed previously in the section containing cross-sectional analysis. That discussion is applicable here.

[16]The reader can verify these results by referring to Tables C13 through C24. Because the F-statistic is simply the ratio of one variance to another and the critical value (3.79) is the same for each ratio, then a random variance from one interval is significantly different from a random variance from another interval, if it is approximately four times larger (or smaller) than the other. For a more detailed discussion, see the introduction to Appendix C.

[17]The reader may recall that this same point was discussed in Chapter IV. However, for the purpose of clarification, it may be well worth explaining in more detail the actual results of this experimentation. We estimated random variances for both larger and shorter intervals (by one and two years) and for intervals of the same length but dated differently. The absolute value of the variances from different intervals did change, but the relative ranking of a state's estimate within each series and the relative ranking of the several series nearly always stayed the same. Because our applications of the variance estimates depended on their relative nature, our applied results were not changed significantly by the choice of an arbitrary set of eight-year intervals. The only exception to these conclusions was that the use of an interval shorter by two years for the computation of random variances did affect the estimates significantly. We feel on safe grounds, though, in not using a six-year interval.

CHAPTER VI

SUMMARY AND CONCLUSIONS

This study has estimated the degree of random variability of prices, yields, and income associated with selected crops and livestock in the United States from 1866 through 1909. These estimates represented objective measures of the amount of uncertainty associated with each series, and they were presented on a state-by-state basis using annual observations. While our estimates were not identical to the traditional concepts of risk or uncertainty, they were, at the least, a reasonable first approximation.

It was found that, in general, the price variability associated with crops was significantly greater than the price variability associated with livestock.[1] A comparison of the yield variability associated with crops and the variability of livestock numbers on farms led to similar conclusions; namely, the estimates associated with crops were significantly greater than those of livestock. The only important exception to this pattern was the price variability of hogs. The variability associated with hog values ranked as high or higher than the corresponding variability of several crops. Income variability estimates applied only to crops.

Regional patterns were shown to exist in the magnitudes of the price, yield, and income variability estimates. The most pronounced regional differences were seen in the price variability estimates, while the income measures had the least pronounced. In general, variability associated with each series was greatest in the less settled and newer agricultural regions of the United States. Estimates for crops grown in the West North Central states and more recently settled regions of the South (for example, Arkansas and Texas), in particular, were higher than those computed for other major agricultural regions. The least variable of all regions were the older areas of the South.

The variability estimates associated with five major crops--corn, cotton, hay, oats, and wheat--grown in the leading producer states were analyzed in detail. The results showed that, in terms of price variability, the corn estimates were by far the highest, while the cotton estimates ranked at the bottom. The magnitudes of price variability for the other three crops, while substantially different than the cotton and corn estimates, were quite close to each other. Yield variabilities fell into two distinct groups. The most variable group consisted of wheat, corn, and cotton, while the least variable group included hay and oats. The wheat crop ranked highest in gross income variability, with cotton being the next most variable. Substantially less than both wheat and cotton were the corn and hay estimates which ranked the lowest in terms of gross income variability.

Because both price and yield variability of corn were relatively high, the most important factor contributing to low gross income variability was the significantly negative year-to-year correlation between the prices and yields of corn. There was only a moderate amount of negative year-to-year correlation between prices and yields of cotton, and the cotton crop had the lowest price variability. Therefore, the most important factor contributing to a relatively high level of gross income variability associated with cotton was relatively high yield variability. For hay and oats, the individual price and yield variabilities were rather good indicators of gross income variabilities. These two crops ranked from moderate to low in terms of all three variability measures. Relatively high yield variability substantially contributed to gross income variability associated with wheat, although moderately high price variability was not unimportant. The year-to-year relationship between wheat prices and yields

had little impact on the high gross income variability estimates, because
it was shown to be insignificant.

The objective measures of agricultural uncertainty computed in
this study were applied to a historical issue concerning the cotton South.
We were interested both in determining whether cotton or corn was a riskier
crop during the postbellum period and in using the estimates to investigate actual acreage decisions. A simple utility maximizing model was
proposed, where the farmers' utility of income was assumed to be a function
of the mean and the random variance of returns and only two crops were
available, namely, cotton and corn. In essence, we investigated the
Wright and Kunreuther (1975) hypothesis which stated that cotton was the
riskier crop and that postbellum Southern farmers displayed gambling
behavior.

Our results showed that, for the region as a whole, cotton had a
higher level of gross income variability than corn. Because farmers
ultimately are interested in the relative variablity of crop returns,
Wright and Kunreuther might be correct in their assertion that, as far
as farmers were concerned, cotton was riskier than corn in the postbellum
South. Yet this was not the case for each major cotton state individually.
Alabama, Georgia, and Mississippi were shown to have had higher levels of
gross income variability for corn than for cotton. In addition, it was
shown that the income variability of cotton was actually one of the lowest
relative to other major alternative crops.

Despite the evidence that cotton, in general, appeared riskier than
corn, the results of the utility maximizing model employed—where we were
interested in the riskiness of the whole portfolio of crops—showed that
the actual acreage choice made by postbellum Southern farmers did not

indicate gambling behavior. On the contrary, the results indicated that postbellum farmers were averse to risk. We computed forty-five estimates which represent states' attitudes toward risk as indicated by their actual acreage decisions. All of these estimates were positive, which implied an aversion to risk, and only three failed to pass a test of significance. In fact, it was argued that the postbellum farmers would have had to alter their crop-mix substantially for it to imply gambling behavior.

In Chapter V of this study, we investigated the relationship between levels of agricultural uncertainty and levels of agrarian discontent during the last third of the nineteenth century in the United States. It has been suggested by some authors (Higgs, 1971; North, 1974) that a major source of farmer discontent was agricultural uncertainty. We tested this assertion for both geographical and temporal relationships.

The results showed that the random variability of the several agricultural series corresponded fairly well with farmer unrest across states. In particular, levels of price variability associated with the major crops of the seventeen-state sample showed a strong positive relationship with levels of protest activity across states. Even though the yield and income variability estimates displayed a weaker relationship, the estimates still supported the contention that uncertainty was a major source of farmer discontent. When the random variability estimates were sub-periodized, so as to more closely correspond with the timing of protest activity, the results indicated an even stronger relationship between agricultural variability and discontent across states.

When the sub-period estimates were used to test for a relationship between temporal patterns in protest activity and temporal patterns in variability, the results were relatively poor. The random variability estimates explained the timing of farmer discontent only four times out of ninety-nine tests. The results improved when a more liberal interpretation of the timing of agrarian unrest was used. Nonetheless, the results still were less than satisfactory. The timing of protest activity was explained in only twenty-six instances out of ninety-nine under weaker conditions.

In common with most empirical studies, this one contained a few problems. Many of these problems were discussed at length in the text while other less serious problems received only a cursory treatment. Our intentions here are to discuss the most general and serious reservations which we have about certain aspects of this study.

Due to data limitations, because our interest was in considering all states, the variability measures associated with individual crops were computed entirely from annual state-wide observations. Yet the variability estimates were intended to represent the amount of agricultural uncertainty faced by an individual farmer. In certain cases, variability measures derived from state data may not have accurately represented fluctuations facing an individual farmer within a state. In particular, this might have been true for variability associated with crop yields. Because of differing regional conditions within a state, some parts of the state may have had quite different amounts of variability than other parts. However, our measures were for an entire state. In addition, part of the state's variability was eliminated, because the yield fluctuations on individual farms were averaged out in compiling the state's data.

It was suggested within the study that state data did not seriously misrepresent prices received by farmers, because of the competitiveness in the market. Thus, the random variability measures, derived from state-wide observations, associated with crop prices were an accurate reflection of the price uncertainty facing an individual farmer.

Other (and possibly more serious) reservations about the general results of this study concern the technique employed for estimating the portion of total variability which could be viewed as unpredictable or random, and the use of these estimates as a risk measure. It has been shown by several authors that no single measure of dispersion of a probability distribution is an appropriate measure of risk, because it will not lead to valid results when applied to all situations. In addition, because our interest was in estimating the levels of agricultural uncertainty from the standpoint of the individual farmer, an implicit assumption of this study was that, in the eyes of the farmer, "random" fluctuations are determined according to the variate difference method.

There is no claim that the risk measures presented in this study are the "best" measures. The sole claim is that they are probably "better" than what has been presented for late nineteenth century agriculture. This is not a difficult proposition to accept when one considers the previous empirical work on this topic. Also, even though the identification of risk with any single measure of dispersion does not lead to universally valid results, a variance-type measure is considered by several scholars to be a reasonable first approximation (Borch, 1968, p. 52; Feldstein, 1969, p. 10) and useful for purposes similar to ours.

Because the primary objective of this study was the estimation of the risk associated with numerous crops and livestock for forty-eight states--necessitating the use of more than 1,800 different time series--the central concern was to find an estimation technique that was both computationally convenient and empirically flexible. Tintner's variate difference method, which requires no a priori specification of rigid functions, appeared to meet our criteria. Although not as robust as more recent developments in time-series analysis (for example, the Box-Jenkins approach), the variate difference method does not require the oftentimes unknown prior information that is needed by the other techniques. Nonetheless, it is realized that a few scholars may have reservations concerning the technique employed.

An additional reservation we have about the results of the study concerns the applications of the random variability estimates contained in Chapters IV and V. Both the issue of acreage choice among postbellum Southern farmers and the issue of agricultural uncertainty's causal role in agrarian discontent were presented in a simple manner. Our reservation is that the analyses presented may be too simple. It appears obvious that risk and expected return are only two among many variables which determine an acreage portfolio, while agricultural risk is only one among many factors influencing farmer unrest. Although the investigation of these issues did not provide definitive answers, it did provide some useful insights for the interpretation of late-nineteenth century agricultural history.

Finally, there exist a number of future research possibilities to which this study could be applied. The most obvious possibilities are that the risk measures presented here could be used in a more fully specified analysis of acreage management and in a more fully specified analysis of the factors influencing agrarian discontent. The variability estimates could be applied to more than simply the problem of acreage management in postbellum Southern agriculture. They could be incorporated into crop selection models for any geographical region during the late nineteenth century.

An additional use of our risk measures might be their incorporation into the explanations of tenure choice provided by Higgs (1973) and Reid (1973). Whether one chose to incorporate the variability estimates into the existing models or to respecify an explanation of tenure choice, the measures presented here might be quite useful. They would be particularly useful, if one agreed that our variability estimates are more accurate measures of agricultural uncertainty than those previously presented. Of course, the use of our measures may not necessarily change the conclusions of the earlier works.

Robert Higgs (1971) suggested that levels of agricultural uncertainty may have played a causal role in the rural-to-urban migration during the late nineteenth century. His argument rested on the assertion that "the uncertainty surrounding farm production was substantially greater than that associated with most types of nonfarm works," and "that farmers migrated to nonfarm jobs seeking greater certainty as well as higher real incomes" (1971, p. 102). Our random variability estimates, along with estimates of the risk associated with earnings from nonfarm occupations, could be used to test this proposition.

Also, the variability estimates presented might be used to investigate further the effects of the agricultural terms of trade on farmer behavior. It has been suggested that the long term trend in prices paid, relative to prices received by farmers, may not be as important as fluctuations in the terms of trade in influencing behavior (Bowman and Keehn, 1974). Thus, the random price variability estimates, which were computed for prices received, might be employed, with appropriate measures of the uncertainty associated with prices paid by farmers, to test this proposition.

FOOTNOTES

[1]Variability here always refers to the <u>random</u> variability coefficient computed by the variate difference method.

BIBLIOGRAPHY

Aldrick, Mark. "Flexible Exchange Rates, Northern Expansion and the Market for Southern Cotton: 1866-1879." Journal of Economic History, vol. 33 (June 1973).

Arrow, Kenneth J. Essays in the Theory of Risk Bearing. Chicago: Markham, 1971.

Barry, Peter J., and Baker, Chester B. "Reservation Prices on Credit Use: A Measure of Response to Uncertainty." American Journal of Economics, vol. 53 (May 1971).

Baumol, William J. "An Expected Gain-Confidence Limit Criterion for Portfolio Selection." Management Science, vol. 9 (October 1963).

Becker, Joseph A., and Harlan, C. L. "Developments in Crop and Livestock Reporting Since 1920." Journal of Farm Economics, vol. 21 (November 1939).

Benedict, Murray R. "Development of Agricultural Statistics in the Bureau of the Census." Journal of Farm Economics, vol. 21 (November 1939).

Bogue, Allan G. Money at Interest: The Farm Mortgage on the Middle Border. Ithaca, N. Y.: Cornell University Press, 1955.

Bogue, Allan G. From Prairie to Corn Belt: Farming on the Illinois and Iowa Prairies in the Nineteenth Century. Chicago: University of Chicago Press, 1963.

Borch, Karl Henrick. The Economics of Uncertainty. Princeton, N. J.: Princeton University Press, 1968.

Borch, Kark Henrick. "A Note on Uncertainty and Indifference Curves." The Review of Economic Studies, vol. 36 (January 1969).

Boussard, Jean-Marc, and Petit, Michel. "Representation of Farmers' Behavior Under Uncertainty with a Focus-Loss Constraint." Journal of Farm Economics, vol. 49 (November 1967).

Bowman, John D. "An Economic Analysis of Midwestern Farm Land Values and Farm Land Income, 1860 to 1900." Yale Economic Essays, vol. 5 (Fall 1965).

Bowman, John D., and Keehn, Richard H. "Agricultural Terms of Trade in Four Midwestern States, 1870-1900." Journal of Economic History, vol. 34 (September 1974).

Brown, William W., and Reynolds, Morgan O. "Debt Peonage Re-examined." Journal of Economic History, vol. 33 (December 1973).

Buck, Solon Justus. The Granger Movement; A Study of Agricultural Organization and Its Political, Economic and Social Manifestations, 1870-1880. Cambridge: Harvard University Press, 1913.

Buck, Solon Justus. The Agrarian Crusade; A Chronicle of the Farmer in Politics. New Haven, Conn.: Yale University Press, 1920.

Bullock, J. Bruce, and Logan, S. H. "A Model for Decision Making Under Uncertainty." Agricultural Economics Research, vol. 21 (October 1969).

Canning, John B. "Rescue Programs and Managed Agricultural Progress." Journal of Farm Economics, vol. 24 (May 1942).

Carstensen, Vernon, ed. Farmer Discontent, 1865-1900. New York: John Wiley and Sons, 1974.

Carter, Harold O., and Dean, Gerald W. "Income, Price, and Yield Variability for Principal California Crops and Cropping Systems." Hilgardia, vol. 30 (October 1960).

Chernoff, Herman, and Moses, Lincoln E. Elementary Decision Theory. New York: John Wiley and Sons, 1959.

Cheung, Steven N. S. "Transaction Costs, Risk Aversion and the Choice of Contractual Arrangements." Journal of Law and Economics, vol. 12 (April 1969).

Chipman, J. S. "The Foundations of Utility." Econometrica, vol. 28 (April 1960).

Cocks, K. D. "Discreate Stochastic Programming." Management Science, vol. 15 (September 1968).

Cooley, Thomas F., and DeCanio, Stephen J. "Rational Expectations in American Agriculture, 1867-1914." Review of Economics and Statistics, vol. 54 (February 1977).

Davis, Lance E.; Hughes, Jonathan R. T.; and McDougall, Duncan M. American Economic History: The Development of a National Economy. 2nd ed. Homewood, Ill.: Richard D. Irwin, 1964.

DeCanio, Stephen. "Cotton 'Overproduction' in Late Nineteenth-Century Southern Agriculture." Journal of Economic History, vol. 33 (September 1974).

DeCanio, Stephen. "Productivity and Income Distribution in the Post-Bellum South." Journal of Economic History, vol. 34 (June 1974).

Ebling, Walter H. "Why the Government Entered the Field of Crop Reporting and Forecasting." Journal of Farm Economics, vol. 21 (November 1939).

Eidman, Vernon R.; Dean, Gerald W.; and Carter, Harold O. "An Application of Statistical Decision Theory to Commercial Turkey Production." Journal of Farm Economics, vol. 49 (November 1967).

Encarnacion, Jose Jr. "Constraints and the Firm Utility Function." Review of Economic Studies, vol. 31 (April 1964).

Fama, E. F. "Portfolio Analysis in a Stable Paretion Market." Management Science, vol. 3 (January 1965).

Farrar, Donald E. The Investment Decision Under Uncertainty. Englewood Cliffs, N. J.: Prentice-Hall, 1962.

Feldstein, Martin. "Mean-Variance Analysis in the Theory of Liquidity Preference and Portfolio Selection." The Review of Economic Studies, vol. 36 (January 1969).

Fisher, Franklin M., and Termin, Peter. "Regional Specialization and the Supply of Wheat in the United States, 1867-1914." Review of Economics and Statistics, vol. 52 (May 1970).

Fisher, Irving. The Nature of Capital and Income. New York: The Macmillan Co., 1906.

Freund, Rudolph J. "The Introduction of Risk into a Programming Model." Econometrica, vol. 24 (July 1956).

Friedman, Milton, and Savage, L. J. "The Expected Utility Hypothesis and the Measurability of Utility." Journal of Political Economy, vol. 60 (December 1952).

Friedman, Milton, and Schwartz, Anna Jacobson. A Monetary History of the United States 1867-1960. Princeton, N. J.: Princeton University Press, 1963.

Gray, Ralph, and Peterson, John M. Economic Development of the United States. Homewood, Ill.: Richard D. Irwin, 1974.

Gould, John P. "Risk, Stochastic Preference, and the Value of Information." Journal of Economic Theory, vol. 8 (May 1974).

Grossman, P. A., and Headley, J. C. Yield and Income Variability for Major Crops in Illinois: A Basis for Farm Decisions. Urbana, Ill.: Department of Agricultural Economics, University of Illinois, 1965.

Hadar, Josef, and Russell, William R. "Rules for Ordering Uncertain Prospects." American Economic Review, vol. 59 (March 1969).

Hadar, Josef, and Russell, William R. "Stochastic Dominance and Diversification." Journal of Economic Theory, vol. 3 (September 1971).

Hadar, Josef, and Russell, William R. "Diversification of Interdependent Prospects." Journal of Economic Theory, vol. 7 (March 1974).

Hale, Roger F. "Estimating Local Market Prices and Farm Labor Since 1920." Journal of Farm Economics, vol. 21 (November 1939).

Halter, Albert N., and Dean, Gerald W. Decision Under Uncertainty with Research Applications. Cincinnati: South-Western, 1971.

Hanoch, G., and Levy, H. "The Efficiency Analysis of Choices Involving Risk." Review of Economic Studies, vol. 36 (July 1969).

Hazell, P. B. R. "A Linear Alternative to Quadratic and Semivariance Programming for Farm Planning Under Uncertainty." American Journal of Agricultural Economics, vol. 53 (February 1971).

Heady, Earl O. "Diversification in Resource Allocation and Minimization of Income Variability." Journal of Farm Economics, vol. 34 (November 1952).

Herstein, I. N., and Milnor, John. "An Axiomatic Approach to Measurable Utility." Econometrica, vol. 21 (April 1953).

Hicks, John D. The Populist Revolt; A History of the Farmers' Alliance and People's Party. Minneapolis: University of Minnesota Press, 1931.

Hicks, John R. "Application of Mathematical Methods to the Theory of Risk" (abstract). Econometrica, vol. 2 (April 1934).

Higgs, Robert. "Railroad Rates and the Populist Uprising." Agricultural History, vol. 44 (July 1970).

Higgs, Robert. The Transformation of the American Economy, 1865-1914: An Essay in Interpretation. New York: John Wiley and Sons, 1971.

Higgs, Robert. "Race, Tenure, and Resource Allocation in Southern Agriculture, 1910." Journal of Economic History, vol. 33 (March 1973).

Higgs, Robert. "Patterns of Farm Rental in the Georgia Cotton Belt, 1880-1900." Journal of Economic History, vol. 34 (June 1974).

Hofstadter, Richard. The Age of Reform; From Bryan to F.D.R. New York: Alfred A. Knopf, 1955.

Jackson, Donald, and Becker, Joseph A. "Revised Estimates of Crop Acreages, New York, 1862-1919." United States Department of Agriculture, Department Circular 373 (April 1926).

Jensen, H. R., and Sundquist, W. B. Resource Productivity and Income for a Sample of West Kentucky Farms. Statistical Bulletin No. 630. Lexington, Ky.: Kentucky Agricultural Experiment Station, 1955.

Jessen, Raymond J. "An Experiment in the Design of Agricultural Surveys." Journal of Farm Economics, vol. 21 (November 1939).

Johnson, Glenn L. "Handling Problems of Risk and Uncertainty in Farm Management Analysis." Journal of Farm Economics, vol. 34 (December 1952).

Kling, William. "Determination of Relative Risks Involved in Growing Truck Crops." Journal of Farm Economics, vol. 24 (August 1942).

Knight, Frank H. Risk, Uncertainty, and Profit. Boston: Houghton Mifflin, 1921.

Kross, Herman E. American Economic Development: The Progress of a Business Civilization. 2nd ed. Englewood Cliffs, N. J.: Prentice-Hall, 1966.

Leland, H. E. "Saving and Uncertainty: The Precautionary Demand for Saving." Quarterly Journal of Economics, vol. 82 (August 1968).

Levhari, D., and Srinivasan, T. N. "Optimal Savings Under Uncertainty." Review of Economic Studies, vol. 36 (April 1969).

Livers, Joe J. "Some Limitations to Use of Coefficient of Variation." Journal of Farm Economics, vol. 24 (November 1942).

Luce, R. Duncan, and Raiffa, Howard. Games and Decisions. New York: John Wiley and Sons, 1958.

McGuire, Robert, and Higgs, Robert. "Cotton, Corn and Risk in the Nineteenth Century: Another View." Explorations in Economic History, vol. 14 (April 1977).

Markowitz, Harry M. "Portfolio Selection." Journal of Finance, vol. 7 (March 1952).

Markowitz, Harry M. Portfolio Selection: Efficient Diversification of Investments. Cowles Foundation Monograph 16. New York: John Wiley and Sons, 1959.

Marschak, Jacob. "Money and the Theory of Assets." Econometrica, vol. 6 (October 1938).

Marschak, Jacob. "Rational Behavior, Uncertain Prospects and Measurable Utility." Econometrica, vol. 18 (April 1950).

Mayhew, Anne. "A Reappraisal of the Causes of Farm Protest in the United States, 1870-1900." Journal of Economic History, vol. 32 (June 1972).

Moore, Charles V. "Income Variability and Farm Size." Agricultural Economics Research, vol. 17 (October 1965).

Moore, Charles V., and Snyder, J. H. "Crop Selection in High-Risk Agriculture." Agricultural Economics Research, vol. 21 (October 1969).

Moore, Henry Ludwell. Economic Cycles: Their Law and Cause. New York: Macmillan, 1914.

Moore, Henry Ludwell. Forecasting the Yield and the Price of Cotton. New York: Macmillan, 1917.

Moore, Henry Ludwell. Generating Economic Cycles. New York: Macmillan, 1923.

Murray, Nat C. "A Close-up View of the Development of Agricultural Statistics from 1900 to 1920." Journal of Farm Economics, vol. 21 (November 1939).

Nugent, Walter T. K. "Some Parameters of Populism." Agricultural History, vol. 50 (October 1966).

North, Douglass C. Growth and Welfare in the American Past. Englewood Cliffs, N. J.: Prentice-Hall, 1966.

North, Douglass C. Growth and Welfare in the American Past. 2nd ed. Englewood Cliffs, N. J.: Prentice-Hall, 1974.

Parker, William N. "Agriculture." In Lance E. Davis et al. American Economic Growth, An Economist's History of the United States. New York: Harper and Row, 1972.

Peterson, John M., and Gray, Ralph. Economic Development of the United States. Homewood, Ill.: Richard D. Irwin, 1969.

Pratt, J. W. "Risk Aversion in the Small and in the Large." Econometrica, vol. 32 (January 1964).

Rae, Allan N. "Stochastic Programming, Utility, and Sequential Decision Problems in Farm Management." American Journal of Agricultural Economics, vol. 53 (August 1971).

Ramsey, Frank P. "Truth and Probability." The Foundations of Mathematics and Other Logical Essays. Edited by R. B. Braithwaite. London: Kegan Paul, 1931.

Ransom, Roger L., and Sutch, Richard. "Debt Peonage in the Cotton South After the Civil War." Journal of Economic History, vol. 32 (September 1972).

Ransom, Roger L., and Sutch, Richard. "The 'Lockin' Mechanism and Over-production of Cotton in the Post-Bellum South." Agricultural History, vol. 49 (April 1975).

Rao, Potluri, and Miller, Roger LeRoy. Applied Econometrics. Belmont, Ca.: Wadsworth, 1971.

Reid, Joseph D. Jr. "Sharecropping as an Understandable Market Response: The Post-Bellum South." Journal of Economic History, vol. 33 (March 1973).

Reid, Joseph D. Jr. "Antebellum Southern Rental Contracts." Explorations in Economic History, vol. 13 (January 1976).

Rothschild, Michael, and Stiglitz, Joseph E. "Increasing Risk I: A Definition." Journal of Economic Theory, vol. 2 (September 1970).

Rothschild, Michael, and Stiglitz, Joseph E. "Increasing Risk II: Its Economic Consequences." Journal of Economic Theory, vol. 3 (March 1971).

Roy, A. D. "Safety First and the Holding of Assets." Econometrica, vol. 20 (July 1952).

Saloutos, Theodore. Farmer Movements in the South, 1865-1933. Berkeley: University of California Press, 1960.

Sarle, Charles F. "Reliability and Adequacy of Farm-Price Data." United States Department of Agriculture, Department Bulletin No. 1480 (March 1927).

Sarle, Charles F. "Adequacy and Reliability of Crop-Yield Estimates." United States Department of Agriculture, Technical Bulletin No. 311 (June 1932).

Sarle, Charles F. "Future Improvement in Agricultural Statistics." Journal of Farm Economics, vol. 21 (November 1939).

Scheffe, Henry. The Analysis of Variance. New York: John Wiley and Sons, 1959.

Schultz, Henry. The Theory and Measurement of Demand. Chicago: University of Chicago Press, 1938.

Scott, John T. Jr., and Baker, Chester B. "A Practical Way to Select an Optimum Farm Plan Under Risk." American Journal of Agricultural Economics, vol. 54, part I (November 1972).

Shannon, Fred A. The Farmer's Last Frontier: Agriculture, 1860-1897. New York: Harper and Row, 1945.

Sharpe, William F. "A Simplified Model for Portfolio Analysis." Management Science, vol. 9 (January 1963).

Sharpe, William F. "Capital Asset Prices: A Theory of Market Equilibrium Under Conditions of Risk." Journal of Finance, vol. 24 (September 1964).

Stiglitz, Joseph E. "The Effects of Wealth, Income and Capital Gains Taxation on Risk Taking." Quarterly Journal of Economics, vol. 83 (May 1969).

Stovall, John G. "Income Variation and Selection of Enterprises." _Journal of Farm Economics_, vol. 48 (December 1966).

Swanson, Earl R. _Variability of Yields and Income from Major Illinois Crops, 1927-1953._ Statistical Bulletin No. 610. Urbana, Ill.: Illinois Agricultural Experiment Station, 1957.

Temin, Peter. "The Causes of Cotton-Price Fluctuations in the 1830's." _Review of Economics and Statistics_, vol. 49 (November 1967).

Thomas, Wayne; Blakeslee, Leroy; Rodgers, LeRoy; and Whittlesey, Norman. "Separable Programming for Considering Risk in Farm Planning." _American Journal of Agricultural Economics_, vol. 54 (May 1972).

Tintner, Gerhard. _The Variate Difference Method._ Cowles Commission for Research in Economics, Monograph No. 5. Bloomington, Ind.: Principia, 1940.

Tintner, Gerhard. _Econometrics._ New York: John Wiley and Sons, 1952.

Tobin, James. "Liquidity Preference as Behavior Towards Risk." _Review of Economic Studies_, vol. 25 (February 1958).

Tobin, James. "Comment on Borch and Feldstein." _The Review of Economic Studies_, vol. 36 (January 1969).

Unger, Irwin. _The Greenback Era: A Social and Political History of American Finance, 1865-1879._ Princeton, N. J.: Princeton University Press, 1964.

United States Department of Agriculture, Agricultural Marketing Service, Crop Reporting Board. _Corn: Acreage, Yield, Production, by States, 1866-1943._ Washington, D. C.: Government Printing Office, June 1954.

United States Department of Agriculture, Agricultural Marketing Service, Crop Reporting Board. _Oats: Acreage, Yield, Production, by States, 1866-1953._ Washington, D. C.: Government Printing Office, June 1954.

United States Department of Agriculture, Agricultural Marketing Service, Crop Reporting Board. _Wheat: Acreage, Yield, Production, by States, 1866-1943._ Statistical Bulletin No. 158. Washington, D. C. Government Printing Office, February 1955.

United States Department of Agriculture, Agricultural Marketing Service, Crop Reporting Board. _Cotton and Cottonseed: Acreage, Yield, Production, Price, Value, by States, 1866-1952._ Statistical Bulletin No. 164. Washington, D. C.: Government Printing Office, June 1955.

United States Department of Agriculture, Agricultural Marketing Service, Crop Reporting Board. Hay: Acreage, Yield, Production, Price, Value, by States, 1866-1953. Statistical Bulletin No. 229. Washington, D. C.: Government Printing Office, June 1958.

United States Department of Agriculture, Agricultural Marketing Service, Crop Reporting Board. Sweet Potatoes: Acreage, Yield, Production, Price, Value, by States, 1868-1953. Statistical Bulletin No. 237. Washington, D. C.: Government Printing Office, September 1958.

United States Department of Agriculture, Agricultural Marketing Service, Crop Reporting Board. Rice, Popcorn, and Buckwheat: Acreage, Yield, Production, Price, Value, by States, 1866-1953. Statistical Bulletin No. 238. Washington, D. C.: Government Printing Office, October 1958.

United States Department of Agriculture, Agricultural Marketing Service, Crop Reporting Board. Barley: Acreage, Yield, Production, Price, Value, by States, 1866-1953. Statistical Bulletin No. 241. Washington, D. C.: Government Printing Office, January 1959.

United States Department of Agriculture, Agricultural Marketing Service, Crop Reporting Board. Potatoes: Acreage, Yield, Production, Price, Value, by States, 1866-1953. Statistical Bulletin No. 251. Washington, D. C.: Government Printing Office, June 1959.

United States Department of Agriculture, Agricultural Marketing Service, Crop Reporting Board. Flaxseed and Rye: Acreage, Yield, Production, Price, Value, by States, 1866-1953. Statistical Bulletin No. 254. Washington, D. C.: Government Printing Office, July 1959.

United States Department of Agriculture, Agricultural Marketing Service, Crop Reporting Board. Tobacco: Acreage, Yield Production, Price, Value, by States, 1866-1965. Statistical Bulletin No. 454. Washington, D. C.: Government Printing Office, June 1970.

United States Department of Agriculture, Bureau of Agricultural Economics. Prices of Farm Products Received by Producers: North Atlantic States. Statistical Bulletin No. 14. Washington, D. C.: Government Printing Office, January 1927.

United States Department of Agriculture, Bureau of Agricultural Economics. Prices of Farm Products Received by Producers: Mountain and Pacific States. Statistical Bulletin No. 17. Washington, D. C.: Government Printing Office, March 1927.

United States Department of Agriculture, Bureau of Agricultural Economics. Prices of Farm Products Received by Producers: North Central States. Statistical Bulletin No. 15. Washington, D. C.: Government Printing Office, May 1927.

United States Department of Agriculture, Bureau of Agricultural Economics. Prices of Farm Products Received by Producers: South Atlantic and South Central States. Statistical Bulletin No. 16. Washington, D. C.: Government Printing Office, June 1927.

United States Department of Agriculture, Bureau of Agricultural Economics. Livestock on Farms, January 1, 1867-1935. Washington, D. C.: Government Printing Office, January 1938.

United States Bureau of the Census. Historical Statistics of the United States, Colonial Times to 1957. Washington, D. C.: Government Printing Office, 1960.

Von Neuman, J., and Morgenstern, O. Theory of Games and Economic Behavior. Princeton, N. J.: Princeton University Press, 1947.

Warren, G. F., and Pearson, F. A. Prices. New York: John Wiley and Sons, 1933.

Wright, Gavin. "Cotton Competition and the Post-Bellum Recovery of the American South." Journal of Economic History, vol. 34 (September 1974).

Wright, Gavin, and Kunreuther, Howard. "Cotton, Corn and Risk in the Nineteenth Century." Journal of Economic History, vol. 35 (September 1975).

Wright, Gavin, and Kunreuther, Howard. "Cotton, Corn, and Risk in the Nineteenth Century: A Reply." Explorations in Economic History, vol. 14 (April 1977).

Yaari, Menahem E. "Convexity in the Theory of Choice Under Risk." Quarterly Journal of Economics, vol. 79 (May 1965).

Yaari, Menahem E. "Some Remarks on Measures of Risk Aversion and on Their Uses." Journal of Economic Theory, vol. 1 (October 1969).

Yamane, Taro. Statistics, An Introductory Analysis. 2nd ed. New York: Harper and Row, 1967.

Yeh, Martin H., and Wu, Roland Y. "Premium Ratemaking in an All Risk Crop Insurance Program." Journal of Farm Economics, vol. 48 (December 1966).

Yule, G. Undy, and Kendall, M. G. An Introduction to the Theory of Statistics. 14th ed. London: Charles Griffin, 1965.

APPENDIX A:

DISCUSSION OF THE DATA AND THE VARIABILITY MEASURES

(TABLES A1 THROUGH A144 INCLUDED)

This appendix includes a complete listing of all data sources; a thorough discussion of the data (including their limitations); a discussion of all adjustments made to the data prior to the estimation of the variability coefficients; a discussion of existing incomplete series; and a presentation of the random variability coefficients for the forty-eight states in Tables A1 through A144. The variability coefficients were estimated for income, yield, and price for the following eleven crops: barley, buckwheat, corn, cotton, oats, potatoes, rye, sweet potatoes, tame hay, tobacco, and wheat. Variability coefficients also were estimated for the number on farms and the value per head of the following six species of livestock: all cattle, hogs, horses, milk cows, mules, and sheep.

Sources

Acreage and yield figures for all crops were from the United States Department of Agriculture, Agricultural Marketing Service, Crop Reporting Board, as follows:

1. Barley: Acreage, Yield, Production, Price, Value, by States, 1866-1953, Statistical Bulletin No. 241 (1959).

2. Corn: Acreage, Yield, Production, by States, 1866-1943, mimeo (1954).

3. Cotton and Cottonseed: Acreage, Yield, Production, Price, Value, by States, 1866-1952, Statistical Bulletin No. 164 (1955).

4. Flaxseed and Rye: Acreage, Yield, Production, Price, Value, by States, 1866-1953, Statistical Bulletin No. 254 (1959).

5. Hay: Acreage, Yield, Production, Price, Value, by States, 1866-1953, Statistical Bulletin No. 229 (1958).

6. Oats: Acreage, Yield, Production, by States, 1866-1943, mimeo (1954).

7. Potatoes: Acreage, Yield, Production, Price, Value, by States, 1866-1953, Statistical Bulletin No. 251 (1959).

8. Rice, Popcorn, and Buckwheat: Acreage, Yield, Production, Price, Value, by States, 1866-1953, Statistical Bulletin No. 238 (1958).

9. Sweet Potatoes: Acreage, Yield Production, Price, Value, by States, 1866-1953, Statistical Bulletin No. 237 (1958).

10. Tobacco: Acreage, Yield, Production, Price, Value, by States, 1866-1965, Statistical Bulletin No. 454 (1970).

11. Wheat: Acreage, Yield, Production, by States, 1866-1943, Statistical Bulletin No. 158 (1955).

Figures for crop prices and figures for the number of livestock on farms and their value per head were taken from the United States Department of Agriculture, Bureau of Agricultural Economics, as follows:

1. Prices of Farm Products Received by Producers: North Atlantic States, Statistical Bulletin No. 14 (1927).

2. Prices of Farm Products Received by Producers: North Central States, Statistical Bulletin No. 15 (1927).

3. Prices of Farm Products Received by Producers: South Atlantic and South Central States, Statistical Bulletin No. 16 (1927).

4. Prices of Farm Products Received by Producers: Mountain and Pacific States, Statistical Bulletin No. 17 (1927).

5. Livestock on Farms, January 1, 1867-1935, mimeo (1938).

For adjustments to some of the price data, which will be outlined later, the following sources were used:

1. United States Bureau of the Census, Historical Statistics of the United States, Colonial Times to 1957 (1960).

2. G. F. Warren and F. A. Pearson, Prices (1933).

Discussion of the Data

All of the U.S.D.A. series used in this study were the revised estimates of the original series. All crop figures were annual state estimates starting in 1866 (or from the first year in which estimates were recorded). Acreage figures are listed as the total acres harvested during the year associated with a given crop. Yield figures are listed as an estimate of the average yield per acre during the year given in units common to the specific crop (for example, bushels or pounds). Price figures are listed as the price per unit, for a given crop, received by the farmer at his local market as of December 1 of each year. In other words, crop prices are an estimated average of the price prevailing for December 1 of each year. The livestock figures collected are estimates of the total number of each species on farms as of January 1 and an estimate of their value per head as of January 1. Both livestock series start in 1867.

The U.S.D.A. estimates of the several series included were based primarily on the reports of voluntary county crop correspondents prior to 1882. Starting in 1882, part-time state reporters, who based their estimates of crop conditions on their own list of voluntary crop correspondents, were appointed. These latter reports were used to supplement the returns from the county correspondents in making the final U.S.D.A. estimate. In 1896, township crop correspondents were added to the list of reporters sending in estimates on crop conditions to the U.S.D.A. Lastly, in 1900, regional field agents were appointed on a full-time basis for the purpose of gathering additional information on crops and livestock (Sarle, 1932, pp. 130-131).

From the beginning in 1866, the reports on acreage, yield, and number of livestock by county correspondents were judgment inquiries, where these volunteer reporters were asked to make an estimate about crop and livestock conditions in their particular locality. Even the estimates furnished by the state statistical agents and regional field agents later still were based on personal observations about a large area in conjunction with the reports supplied by their own crop correspondents using judgment inquiries (Sarle, 1932, p. 8). In fact, "judgment data were the chief source of information used in preparing reports prior to 1920" (Becker and Harlan, 1939, p. 800). Yet some individual-farm samples (where a crop reporter or other farmer is asked to report on the actual acreage, yield, or number of livestock on his farm) were used as early as the 1880's. These "estimates of the yields on a large number of individual farms were obtained for the first time and were used as a check on the other sources of information" (Sarle, 1932, p. 130). Also, many other sources of information on acreage and yields were made available by the addition of the state statistical agents--for example, returns gathered from threshers, operators of mills, and grain elevator companies. According to Sarle, these "additional reports obtained by state agents probably doubled the size of the sample for most states" (1932, p. 130).

The several types of crop reporters--county correspondents, township correspondents, and state agents--all received schedules containing the same questions from the U.S.D.A. (Murray, 1939, p. 712). In the beginning of the reports, the inquiry was made about the acreage, yield, and livestock numbers as a percentage of the previous year (Sarle, 1932, p. 132; Becker and Harlan, 1939, p. 800; Fisher and Temin, 1970, p. 136). The U.S.D.A. used these several reports to estimate percentage changes from year to year, which were then used with census data (as benchmarks) to estimate the various original series on an annual basis. Sarle (1939, pp. 7-8) mentioned that the inquiries, at times, included questions about the actual number instead of simply percentage changes. However, there was no clarification as to how the data were compiled and utilized, nor when this method was used.

Because the original series were based primarily on a sampling and estimating procedure (using percentage changes) built upon an enumeration base (the census year), the original annual estimates of various series quite often had to be adjusted drastically to bring them in line with a new census enumeration (Jackson and Becker, 1926, p. 5). When new census figures became available, no attempt was made to revise the estimates of the earlier years. Thus, there exist sharp changes in the figures around each census year, and there appears to be no explanation of the procedure followed in making these adjustments (Jackson and Becker, 1926, pp. 2-3).

The procedure for revising the annual data estimated from the original series was similar for all states. Because the original estimates for the various series were calculated from base years (census enumerations), it was essential that the Department of Agriculture had accurate figures on these benchmark years. They were considered accurate for most crops, and inaccurate for most livestock species. Independent sources of information were necessary for the revisions. For example, state assessment records were used as a check on the original estimates of livestock numbers (United States Department of Agriculture, 1938, p. 5).

Crop acreages were revised by assuming that the census enumerations were reasonably accurate for most crops, and thus, the adjustments were made in the trend between census years on the basis of the percentage changes given in the various original reports. That is, the Department of Agriculture statisticians, through their analysis, determined that the original year-to-year changes were accurate and only the trends needed changing (Jackson and Becker, 1927, pp. 2-5).

The livestock numbers were a different story. Livestock totals also were estimated by applying percentage changes, supplied by crop reporters, to the census enumerations so as to calculate an annual figure. However, because of conflicting instructions to enumerators, incomplete count of range animals, changing classifications of animals, and inconsistent enumeration dates, the census figures were considered to be a very poor base upon which to estimate annual numbers. Even the original estimates of the percentage changes from year to year were considered of limited value (United States Department of Agriculture, 1938, pp. 2-4). Thus, using state records on the assessment of livestock, which the Department of Agriculture determined were dependable checks on both trend and year-to-year changes in livestock numbers, the revisions were made.

It is interesting to note that the original crop-yield estimates did not have nearly as many inaccuracies as many of the other annual estimates. In his extensive study of the relationship between the original Department of Agriculture estimates and census figures, Sarle concluded that for most major crops the Department "estimates of yield per acre check sufficiently well with the yields derived from census data" except possibly for some crops in the census years 1879 and 1889 (1932, p. 128). This conclusion was based on his examination of the sampling and weighting procedures used by the Department in arriving at the estimates. Given that the Department used several reports from various correspondents-- county, township, state agents, and regional agents--as cross checks and also used several weighting schemes, Sarle was of the opinion that for "the more important crops in the larger states a sample of this size stratified by counties would be fairly adequate in size, and differences in yield per acre for a given crop from year to year would be reasonably significant from a statistical standpoint" (1932, p. 130). Thus, only minor modifications were made in the crop-yield data "according to a study of the usual

relation between reported and census yields in other census years"
(Jackson and Becker, 1926, p. 3).

The revised estimated values per head of livestock were, for the
most part, "the original values as represented in the years to which they
apply. Some changes were made in the original figures for different
states in some years for some species when the original value in relation
to preceding or following years, or in relation to values in nearby states
seemed to be out of line. Such adjustments, however, were relatively
few" (United States Department of Agriculture; 1938, p. 17).

Crop prices, as revised, were for the most part the original values
reported. The December 1 prices reflected annual changes quite well;
however, because they are annual changes as of December 1, they did not
necessarily indicate the average level of price during the marketing
season (Hale, 1939, pp. 828-830). Because these estimates are based on a
sufficient number of reports from both surplus and deficit producing areas
within a state and are averaged within crop reporting districts (usually
nine in a state) and then a weighted average of districts (according to
the number of acres of the crop within the district) is taken to determine
the state estimate, these estimates are considered quite reasonable and
reliable (Sarle, 1927, p. 64). Even the prices of minor products are
considered reliable for studying year-to-year movements (United States
Department of Agriculture, January 1927, pp. 2-3).

Three additional observations concerning the farm-price data have
been found in the literature. These points have not been commonly
mentioned in historical studies using the figures. In his extensive study
on the reliability of farm-price data, Sarle described these characteristics as:

1. Collection of farm prices began in 1867 (not 1866), when farm prices were reported as of January 1. In 1872, the reporting date was changed to December 1. After the change, it became customary to consider the 1867 to 1872 prices as December 1 prices. Thus, they now are published as December 1, 1866-1871, making a complete series of December 1 prices.

2. All crop prices and livestock values for the period 1866-1878 were published in gold equivalent prices. That is, the actual prices (in dollars) of crops and livestock were reduced to a gold basis for this period only.

3. The crop prices, as reported, were estimates of the average price of all grades and classes of commodities being sold in the local market. Thus, there is no way to distinguish between the effects of the quality of the crop on the price estimate and the effects of other factors.

It should be noted that the prices of crops reported in the several bulletins of the Department of Agriculture in the 1950's no longer are reduced to a gold basis between 1866 and 1878. These price figures are all listed in current dollars of the years to which they apply.

How accurate are the revised estimates for the purpose at hand? As far as crop yields, crop prices, and livestock values are concerned, the consensus opinion among Department statisticians is that the figures now available are a reliable indication of the actual general trend and changes in the variables from one year to the next (Sarle, 1927, 1932; United States Department of Agriculture, January 1927; Hale, 1939). Even though the estimates cannot be considered accurate as to the absolute number in a given year, they appear to be accurate for our purposes. However, it must be kept in mind that we are dealing with averages; thus, they can never be considered the actual figure facing any given individual farmer. Nonetheless, because we are interested in measuring the variability of the several series for a given crop in a given state facing farmers in general, we do not feel that the available data create any major problems.

For acreage totals and livestock numbers, the accuracy of the figures is more questionable. Even though the following quote refers only to livestock numbers, it is a reasonable description of the opinion of the Department statisticians toward both estimates:

> From what has been indicated as to the character of information available for making these revisions, it is obvious that no high degree of accuracy can be claimed for the absolute numbers for a given year, either by states or by groups of states . . . the numbers reported can always be considered as below the actual for the date . . . it follows that the poor enumerations must have given figures considerably below the actual. How much below cannot be determined . . . the revised figures for the years before 1900, both in relative accuracy and in indicating trends, undoubtedly are less dependable than those for the years since that date (United States Department of Agriculture, January 1938, p. 6).

It should be fairly clear that we cannot place much faith in the absolute value of any of the individual figures. The sampling procedures used in the nineteenth century, no matter how representative for then, are not what one would consider the "best" technique. Subjective judgment reports are always going to be "subject to the personal bias of the reporter" and some items always will be left out (Sarle, 1939, p. 841). No matter how careful the reporters are, they can never gauge the actual changes in crop or livestock conditions without error (Becker and Harlan, 1939, p. 801). However, most of the bias inherent in this type of estimating procedure could be expected to affect different crops and livestock within a state and maybe among different states in a similar manner

(Sarle, 1932, pp. 128-136). Another problem with the data is that as the figures are averages, for the most part, for a state, they always will mask any differences in values that exist for a variable within a state. Because heterogeneity of the several series did exist within states (Jessen, 1939), one must be aware of this problem when drawing conclusions.

Taking the above characteristics of the data into consideration, we still are of the opinion that for the study at hand the available figures will serve our purpose well. In other words, because we are interested, in general, solely in the relative nature of our estimates, our conclusion is that the various U.S.D.A. series used here are accurate with respect to their relative levels.

Adjustments to Raw Data

Both major and minor adjustments to the data were performed before our variability measures were calculated. Not all series were adjusted. The major adjustments made were to crop and livestock price estimates for the years 1866 to 1878, to cotton prices from 1869 to 1881, and to sweet potato prices prior to 1899. Minor adjustments were made to different crop prices in several states for various years.

Because crop prices and livestock values for the years 1866-1878, as published in 1927, were reduced to a gold basis, while all prices and values after 1878 were listed in current dollars, it should be apparent that the estimated crop prices and livestock values for the earlier years are not comparable to the same estimates in later years. The resumption of specie payments and the return to a gold standard in 1879 were not sufficient justifications for estimating prices and values prior to that date in terms of the gold value of currency. The Department of Agriculture and scholars still using the 1927 published figures are suffering from an illusion about the gold standard. Simply because greenbacks could be turned in for gold after 1878, did not mean that the value of gold measured in dollars remained constant throughout the nineteenth century. In fact, the gold value of currency fluctuated during the entire period (Warren and Pearson, 1933, p. 351). Thus, it is not correct to reduce the figures between 1866 and 1878 to a gold basis and leave the later figures in current dollars. It is interesting to note that the Department of Agriculture must have realized its error. The bulletins published in the 1950's, which contained crop prices, all used current dollar estimates.

Because price data in current dollars were available for several crops, all crop prices were not adjusted. The estimates of December 1 prices contained in the bulletins published in the 1950's were used for those crops for which they existed. The crop prices still listed on a gold basis in 1866-1878--corn, hay, oats, and wheat--and all livestock values, which were still on a gold basis for those years, were adjusted upward to current dollars in the years to which they apply. The December estimates of the gold value of currency for each year and the January estimates for each year, contained in Warren and Pearson (1933, p. 351), were used as the adjustment index for crop prices and livestock values, respectively.

The annual state series of cotton prices did not start until 1876 and contained missing observations in 1877 and 1881. We adjusted the cotton prices series by using an average United States December 1 cotton price (United States Bureau of the Census, 1960, p. 302) for the years between 1869 and 1875 and for 1877 and 1881 in place of the nonexistent state prices. This should not create any major problems. Cooley and

DeCanio argued:

> The correlation between the U.S. prices and the state prices was
> quite high after 1882 because of the competitiveness of the national
> cotton market, and since this market was well developed before 1882
> it is unlikely that any substantial error is introduced by use of the
> national price for the early years of the sample (1977, p. 11, note 4).

Thus, we should not expect problems for the whole period under study;
however, estimates of cotton price variability for the earlier years (1869-
1875) would obviously be the same in all states.

The sweet potato price series was adjusted because of missing observations in every state for which estimates existed. The series started in
1868 and had missing observations in 1876-1878, 1881, 1892, 1893, and
1898. Because there was not a competitive national market for sweet potatoes, an average U.S. price could not be used to replace nonexistent state
prices. Therefore, a straight linear interpolation between existing figures
for each state was used to estimate and replace the missing observations.
However, using linear interpolation for three years (1976-1878) did not
appear to be appropriate. Thus, all observations prior to 1879 were dropped
and interpolation was used only for the latter missing observations--1881,
1892, 1893, and 1898.

Minor adjustments were made to several price series because of
missing observations. Because the figure for only one year was missing
for the most part, we chose to use a straight linear interpolation between
the existing figures (before and after the missing figure) to estimate and
replace the nonexistent number. In a few cases, two consecutive observations were missing. They were of such minor importance, however, that the
use of linear interpolation would not create significant bias. All inter-
polations made are listed below:

1. Price of buckwheat in North Dakota for 1886.

2. Price of buckwheat in South Dakota for 1886.

3. Price of buckwheat in Delaware for 1887 and 1888.

4. Price of cotton in Virginia for 1894.

5. Price of barley in West Virginia for 1880.

6. Price of barley in North Carolina for 1888.

7. Price of barley in Georgia for 1888.

8. Price of buckwheat in Kentucky for 1887 and 1888.

9. Price of barley in Alabama for 1885, 1887, and 1888.

10. Price of rye in Mississippi in 1885.

11. Price of tobacco in Mississippi in 1909.

12. Price of oats in Louisiana for 1876 and 1877.

13. Price of rye in Louisiana for 1885 and 1888.

Incomplete Series

Several crop and livestock series for various states are either nonexistent or incomplete. The Department of Agriculture did not publish estimates for any crop or livestock for which there were too few observations. Also, the estimates were not published for a particular year for a given crop or livestock series, if there were too few observations for that year. Thus, the lack of variability measures in our tables for any particular series indicates that the series did not exist or that it had only scattered observations (that is, it was incomplete). Also, variability measures are included in our tables for some incomplete series with short time periods, when enough consecutive observations existed to allow the calculations.

The data for all series for certain states are incomplete because of lack of information by the Department of Agriculture. Each state whose recorded histories of agricultural statistics began later than 1866 is listed below (the date in parentheses is that state's beginning date):

Minnesota (1867)	Colorado (1880)
North Dakota (1882)	New Mexico (1882)
South Dakota (1882)	Arizona (1882)
West Virginia (1867)	Utah (1882)
Oklahoma (1893)	Nevada (1870)
Montana (1882)	Washington (1882)
Idaho (1882)	Oregon (1869)
Wyoming (1882)	California (1868)

Nearly every livestock series for the above states was extended backwards to 1867 by using state assessment records by the Department of Agriculture in their published estimates. However, because no original Department estimates existed to check these extrapolations against and because the various crop series of these states began in the years listed above, we have chosen to use only the revised livestock figures starting in the year shown above in our calculations.

Explanation of Tables A1 through A144

The tables included in this appendix, A1 through A144, present our estimates of the mean, random standard deviation, and random variability coefficient of prices, yields, and gross income of the crops in forty-eight states in which historical agricultural statistics exist. Because the most complete price and yield series exist for wheat, corn, oats, barley, rye, buckwheat, potatoes, sweet potatoes, tame hay, cotton and tobacco, these are the eleven crops for which estimates are made. Historical livestock series exist for horses, mules, sheep, hogs, all cattle, and milk cows. Thus, we have estimated the mean, random standard deviation, and random variability coefficient of the value per head and number of livestock on farms for forty-eight states. These estimates also are presented in Tables A1 through A144.

The state's estimates of the various series are presented in our tables by census regions in the following order: North Atlantic, East North Central, West North Central, South Atlantic, South Central, and Western.

Because a number of our estimates have a particular common characteristic, alphabetic superscripts have been used throughout the tables to indicate the specific characteristic which applies to that estimate. The explanations of these superscripts are listed below:

(a) This estimate was based on a random variance which did not remain perfectly stable according to our choice criterion throughout the whole differencing process. However, after three or four successive differences the estimated random variance was reasonably stable.

(b) This estimate was based on a random variance which was never stable according to our choice criterion throughout the whole differencing process. Nevertheless, most of the systematic component of this series was removed by the fourth successive difference, thus yielding a valid estimate of the random variance.

(c) This estimate was based on a very short series which prevented the use of our choice criterion. Thus, the researcher must use his own judgment as to what constitutes a valid estimate of the random variance.

(d) All estimates of price and income variability of sweet potatoes were based on thirty-one observations from 1879 to 1909, unless otherwise noted.

(e) All estimates of yield variability of sweet potatoes were based on forty-two observations from 1868 to 1909, unless otherwise noted.

    (f) All estimates of price and income variability of cotton were
based on forty-one observations from 1869 to 1909, unless
otherwise noted.

Any other estimates which are based on fewer than forty-four observations
(from 1866 to 1909) are indicated by a numerical superscript. These
superscripts are explained at the bottom of each table which contains the
estimate.

Selected Crops and Livestock, Maine: Ranking
By Price Random Variability Coefficients

	Random Variability Coefficient	Standard Deviation	Mean	Unit
	(Percent)			
Crop				
Potatoes	24.79	.233	.94	$/cwt.
Tame Hay	14.41	1.725	11.97	$/ton
Oats	10.68	.052	.49	$/bu.
Buckwheat	8.94	.055	.61	$/bu.
Rye	8.57	.087	1.02	$/bu.
Corn	7.75	.062	.80	$/bu.
Wheat	7.20	.096	1.33	$/bu.
Barley	4.69	.036	.76	$/bu.
Livestock				
Hogs	10.33	1.001	9.69	$/Head
All Cattle	8.69	2.517	28.96	$/Head
Milk Cows	7.96	2.362	30.92	$/Head
Sheep	7.51	.228	3.04	$/Head
Horses	2.42	1.964	81.14	$/Head

TABLE A2

Selected Crops and Livestock, Maine: Ranking By
Yield and Number on Farms Random Variability Coefficients

	Random Variability Coefficient	Standard Deviation	Mean	Unit
	(Percent)			
Crop				
Potatoes	20.07	15.495	77.21	Cwt./acre
Rye	15.32	1.874	12.23	Bu./acre
Corn	15.16	5.097	33.61	Bu./acre
Oats	14.28	4.364	30.57	Bu./acre
Tame Hay	12.10	.111	.92	Ton/acre
Buckwheat	11.43	2.281	19.96	Bu./acre
Wheat	8.78	1.540	17.53	Bu./acre
Barley	3.67	.891	24.28	Bu./acre
Livestock				
Sheep	1.85	7.071	383.0	1000's of Head
Milk Cows	1.35	2.091	154.4	1000's of Head
Hogs	1.33	.979	73.5	1000's of Head
All Cattle	1.03[b]	3.362	326.6	1000's of Head
Horses	.67	.655	98.1	1000's of Head

TABLE A3

Selected Crops, Maine: Ranking By
Gross Income Random Variability Coefficients

Crop	Random Variability Coefficient	Standard Deviation	Mean	Unit
	(Percent)			
Potatoes	19.15	13.409	70.04	$/acre
Corn	16.81	4.444	26.43	$/acre
Rye	14.85	1.839	12.38	$/acre
Buckwheat	12.13	1.464	12.07	$/acre
Oats	12.02	1.774	14.75	$/acre
Wheat	9.56	2.106	22.04	$/acre
Tame Hay	9.54	1.028	10.78	$/acre
Barley	4.99	.905	18.15	$/acre

TABLE A4

Selected Crops and Livestock, New Hampshire: Ranking
By Price Random Variability Coefficients

	Random Variability Coefficient	Standard Deviation	Mean	Unit
	(Percent)			
Crop				
Potatoes	25.62	.257	1.00	$/cwt.
Tame Hay	12.95	1.702	13.42	$/ton
Buckwheat	11.30	.074	.66	$/bu.
Oats	10.10	.051	.50	$/bu.
Barley	9.21	.074	.81	$/bu.
Corn	9.15	.072	.79	$/bu.
Wheat	6.75	.094	1.40	$/bu.
Rye	6.62	.064	.97	$/bu.
Livestock				
Hogs	7.11	.801	11.27	$/Head
Horses	6.09	4.636	76.10	$/Head
Sheep	5.32	.158	2.96	$/Head
All Cattle	5.13	1.540	30.04	$/Head
Milk Cows	4.65	1.518	32.63	$/Head

Selected Crops and Livestock, New Hampshire: Ranking By
Yield and Number of Farms Random Variability Coefficients

	Random Variability Coefficient	Standard Deviation	Mean	Unit
	(Percent)			
Crop				
Rye	16.63	1.869	11.23	Bu./acre
Potatoes	16.49	10.957	66.43	Cwt./acre
Corn	12.15	4.660	38.36	Bu./acre
Oats	9.01	3.086	34.26	Bu./acre
Wheat	8.74	1.411	16.13	Bu./acre
Buckwheat	8.48	1.869	22.04	Bu./acre
Tame Hay	8.35	.081	.97	Ton/acre
Barley	6.46	1.577	24.41	Bu./acre
Livestock				
Sheep	2.07	3.362	162.5	1000's of Head
Hogs	1.36	.623	46.0	1000's of Head
Milk Cows	1.01	1.078	106.7	1000's of Head
All Cattle	1.00	2.240	225.0	1000's of Head
Horses	.75	.376	49.1	1000's of Head

TABLE A6

Selected Crops for New Hampshire: Ranking By
Gross Income Random Variability Coefficients

Crop	Random Variability Coefficient	Standard Deviation	Mean	Unit
	(Percent)			
Rye	23.06	2.504	10.86	$/acre
Potatoes	20.02	12.957	64.71	$/acre
Corn	15.16	4.562	30.10	$/acre
Buckwheat	13.60	1.963	14.43	$/acre
Tame Hay	11.40	1.448	12.70	$/acre
Oats	10.54	1.790	16.98	$/acre
Wheat	9.65	2.180	22.60	$/acre
Barley	6.78	1.313	19.37	$/acre

Selected Crops and Livestock, Vermont: Ranking By
By Price Random Variability Coefficients

	Random Variability Coefficient	Standard Deviation	Mean	Unit
	(Percent)			
Crop				
Potatoes	26.16	.218	.83	$/cwt.
Barley	10.67	.082	.77	$/bu.
Tame Hay	10.29	1.132	11.01	$/ton
Oats	9.64	.045	.46	$/bu.
Wheat	9.56	.119	1.24	$/bu.
Rye	9.22	.079	.86	$/bu.
Corn	9.04	.069	.77	$/bu.
Buckwheat	6.43	.040	.62	$/bu.
Livestock				
Hogs	11.78	1.106	9.39	$/Head
Sheep	9.61	.315	3.28	$/Head
All Cattle	6.58	1.858	28.24	$/Head
Horses	5.11	3.884	75.95	$/Head
Milk Cows	4.68	1.439	30.76	$/Head

TABLE A8

Selected Crops and Livestock, Vermont: Ranking By
Yield and Number on Farms Random Variability Coefficients

	Random Variability Coefficient	Standard Deviation	Mean	Unit
	(Percent)			
Crop				
Potatoes	16.19	10.761	64.48	Cwt./acre
Buckwheat	12.76	2.550	19.99	Bu./acre
Corn	12.40	4.901	39.52	Bu./acre
Rye	11.53	1.641	14.23	Bu./acre
Wheat	9.29	1.701	18.31	Bu./acre
Barley	8.21	2.265	27.57	Bu./acre
Tame Hay	7.84	.089	1.13	Ton/acre
Oats	7.41	2.564	34.59	Bu./acre
Livestock				
Sheep	1.36	4.971	366.5	1000's of Head
Hogs	1.21	.913	75.7	1000's of Head
All Cattle	.63	2.638	418.7	1000's of Head
Milk Cows	.56	1.343	238.6	1000's of Head
Horses	.33	.268	81.2	1000's of Head

Selected Crops, Vermont: Ranking By
Gross Income Random Variability Coefficients

Crop	Random Variability Coefficient	Standard Deviation	Mean	Unit
	(Percent)			
Potatoes	21.11	11.416	54.08	$/acre
Rye	15.83	1.965	12.41	$/acre
Barley	13.09	2.706	20.68	$/acre
Buckwheat	12.06	1.475	12.23	$/acre
Corn	11.65	3.503	30.08	$/acre
Wheat	10.51	2.345	22.33	$/acre
Oats	9.89	1.574	15.92	$/acre
Tame Hay	8.62	1.064	12.35	$/acre

TABLE A10

Selected Crops and Livestock for Massachusetts: Ranking
By Price Random Variability Coefficients

	Random Variability Coefficient	Standard Deviation	Mean	Unit
	(Percent)			
Crop				
Tobacco	21.57	.035	.16	$/lb.
Potatoes	20.86	.254	1.22	$/cwt.
Barley	11.40	.101	.88	$/bu.
Oats	11.20	.058	.52	$/bu.
Wheat	10.41	.166	1.60	$/bu.
Corn	8.86	.069	.78	$/bu.
Buckwheat	8.80	.066	.75	$/bu.
Tame Hay	8.77	1.574	17.96	$/ton
Rye	8.66	.078	.90	$/bu.
Livestock				
Hogs	11.50	1.360	11.83	$/Head
Sheep	9.20	.335	3.65	$/Head
Milk Cows	8.04[a]	3.208	39.89	$/Head
All Cattle	7.26	2.652	36.51	$/Head
Horses	4.93	4.819	97.79	$/Head

Selected Crops and Livestock, Massachusetts: Ranking By
Yield and Number on Farms Random Variability Coefficients

	Random Variability Coefficient	Standard Deviation	Mean	Unit
	(Percent)			
Crop				
Potatoes	16.53	9.345	56.55	Cwt./acre
Corn	10.29	3.833	37.26	Bu./acre
Buckwheat	9.52	1.306	13.71	Bu./acre
Barley	8.84	2.106	23.82	Bu./acre
Oats	7.84[a]	2.517	32.12	Bu./acre
Tame Hay	6.84	.079	1.16	Ton/acre
Tobacco	5.72	85.079	1488.16	Lb./acre
Rye	5.09	.642	12.59	Bu./acre
Livestock				
Sheep	1.78	1.034	58.0	1000's of Head
Hogs	1.04	.802	77.4	1000's of Head
Horses	.74	.458	62.0	1000's of Head
All Cattle	.70	1.878	267.3	1000's of Head
Milk Cows	.60	.982	164.4	1000's of Head

TABLE A12

Selected Crops for Massachusetts: Ranking By
Gross Income Random Variability Coefficients

Crop	Random Variability Coefficient	Standard Deviation	Mean	Unit
	(Percent)			
Tobacco	21.41	50.860	237.52	$/acre
Potatoes	19.69	13.356	67.82	$/acre
Barley	13.51	2.892	21.41	$/acre
Rye	11.91[a]	1.317	11.06	$/acre
Oats	11.58	1.906	16.46	$/acre
Buckwheat	10.76	1.085	10.09	$/acre
Corn	9.73[a]	2.780	28.57	$/acre
Tame Hay	6.61	1.353	20.48	$/acre

Selected Crops and Livestock for Rhode Island: Ranking
By Price Random Variability Coefficients

	Random Variability Coefficient	Standard Deviation	Mean	Unit
	(Percent)			
Crop				
Potatoes	21.61	.278	1.29	$/cwt.
Rye	13.31	.132	.99	$/bu.
Corn	10.01	.084	.84	$/bu.
Oats	9.94	.051	.51	$/bu.
Barley	7.93	.070	.89	$/bu.
Tame Hay	5.44	1.046	19.23	$/ton
Livestock				
Sheep	11.71	.449	3.83	$/Head
Hogs	7.16	.802	11.21	$/Head
Horses	7.05	6.873	97.50	$/Head
All Cattle	5.51	2.040	37.04	$/Head
Milk Cows	5.32	2.053	38.61	$/Head

TABLE A14

Selected Crops and Livestock, Rhode Island: Ranking By
Yield and Number on Farms Random Variability Coefficients

	Random Variability Coefficient	Standard Deviation	Mean	Unit
	(Percent)			
Crop				
Potatoes	22.40	13.580	60.64	Cwt./acre
Rye	11.70	1.688	14.43	Bu./acre
Tame Hay	10.08	.106	1.05	Ton/acre
Barley	7.48	1.668	22.30	Bu./acre
Oats	6.09[a]	1.815	29.78	Bu./acre
Corn	5.29	1.731	32.73	Bu./acre
Livestock				
Hogs	2.78	.366	13.2	1000's of Head
Horses	2.43	.244	10.0	1000's of Head
Sheep	2.14	.315	14.7	1000's of Head
All Cattle	1.57	.563	35.9	1000's of Head
Milk Cows	1.09	.236	21.5	1000's of Head

Selected Crops for Rhode Island: Ranking By
Gross Income Random Variability Coefficients

Crop	Random Variability Coefficient	Standard Deviation	Mean	Unit
	(Percent)			
Potatoes	16.40	12.642	77.08	$/acre
Rye	14.46	2.077	14.36	$/acre
Tame Hay	10.84	2.172	20.03	$/acre
Corn	10.18	2.760	27.12	$/acre
Barley	9.20	1.963	21.33	$/acre
Oats	9.07	1.372	15.13	$/acre

TABLE A16

Selected Crops and Livestock for Connecticut: Ranking
By Price Random Variability Coefficients

	Random Variability Coefficient	Standard Deviation	Mean	Unit
	(Percent)			
Crop				
Potatoes	22.94	.281	1.23	$/cwt.
Tobacco	17.61	.030	.17	$/lb.
Wheat	12.60	.177	1.41	$/bu.
Barley	12.24	.110	.90	$/bu.
Oats	11.86	.059	.50	$/bu.
Tame Hay	10.03	1.713	17.07	$/ton
Buckwheat	7.92	.061	.78	$/bu.
Corn	7.82	.061	.78	$/bu.
Rye	5.79	.051	.87	$/bu.
Livestock				
Hogs	10.22	1.181	11.55	$/Head
Sheep	8.04	.320	3.98	$/Head
Horses	4.46	4.035	90.39	$/Head
Milk Cows	3.95	1.484	37.56	$/Head
All Cattle	2.76	.968	35.07	$/Head

Selected Crops and Livestock, Connecticut: Ranking By
Yield and Number on Farms Random Variability Coefficients

	Random Variability Coefficient	Standard Deviation	Mean	Unit
	(Percent)			
Crop				
Potatoes	19.84	11.289	56.91	Cwt./acre
Tame Hay	13.34	.144	1.08	Ton/acre
Corn	12.17	4.322	35.50	Bu./acre
Buckwheat	8.97	1.264	14.08	Bu./acre
Oats	8.52	2.596	30.15	Bu./acre
Wheat	6.69	1.145	17.12	Bu./acre
Rye	5.22	.796	15.24	Bu./acre
Tobacco	3.68	58.151	1580.10	Lb./acre
Barley	2.08	.455	21.83	Bu./acre
Livestock				
Hogs	1.41	.736	52.3	1000's of Head
Sheep	1.38	.657	47.5	1000's of Head
All Cattle	.93	2.010	226.1	1000's of Head
Milk Cows	.79	.953	120.6	1000's of Head
Horses	.54	.245	45.0	1000's of Head

TABLE A18

Selected Crops for Connecticut: Ranking By
Gross Income Random Variability Coefficients

Crop	Random Variability Coefficient	Standard Deviation	Mean	Unit
	(Percent)			
Tobacco	20.34[a]	55.046	270.60	$/acre
Wheat	14.27	3.591	25.16	$/acre
Potatoes	12.23[a]	8.303	67.86	$/acre
Buckwheat	12.21	1.313	10.75	$/acre
Oats	11.83	1.776	15.01	$/acre
Barley	11.79	2.364	20.05	$/acre
Corn	9.80	2.682	27.35	$/acre
Tame Hay	9.69	1.741	18.08	$/acre
Rye	5.91	.769	13.02	$/acre

Selected Crops and Livestock, New York: Ranking
By Price Random Variability Coefficients

	Random Variability Coefficient	Standard Deviation	Mean	Unit
	(Percent)			
Crop				
Potatoes	28.83	.264	.92	$/cwt.
Tobacco	18.24	.020	.11	$/lb.
Oats	15.38	.065	.42	$/bu.
Barley	11.60	.087	.75	$/bu.
Buckwheat	11.13	.072	.65	$/bu.
Wheat	11.11	.125	1.13	$/bu.
Tame Hay	10.23	1.254	12.26	$/ton
Corn	9.24	.062	.67	$/bu.
Rye	7.36	.055	.75	$/bu.
Livestock				
Hogs	7.07	.620	8.77	$/Head
Milk Cows	6.19	2.115	34.19	$/Head
All Cattle	5.89	1.883	32.00	$/Head
Sheep	5.82	.215	3.70	$/Head
Horses	3.10	2.736	88.12	$/Head
Mules	3.01	2.980	98.86	$/Head

Selected Crops and Livestock, New York: Ranking By
Yield and Number on Farms Random Variability Coefficients

	Random Variability Coefficient	Standard Deviation	Mean	Unit
	(Percent)			
Crop				
Potatoes	17.08	9.210	53.91	Cwt./acre
Wheat	16.55	2.900	17.52	Bu./acre
Oats	12.87	3.798	29.51	Bu./acre
Buckwheat	12.67	2.012	15.88	Bu./acre
Barley	12.11	2.800	23.11	Bu./acre
Tame Hay	10.72	.124	1.15	Ton/acre
Tobacco	10.23	119.973	1172.50	Lb./acre
Rye	9.39	1.228	13.07	Bu./acre
Corn	9.08	2.912	32.07	Bu./acre
Livestock				
Mules	4.48	.178	4.0	1000's of Head
Sheep	1.78	26.472	1486.4	1000's of Head
Hogs	1.34	8.843	658.5	1000's of Head
All Cattle	.81	19.033	2358.6	1000's of Head
Horses	.28	1.750	619.9	1000's of Head
Milk Cows	.17	2.508	1455.5	1000's of Head

TABLE A21

Selected Crops, New York: Ranking By
Gross Income Random Variability Coefficients

Crop	Random Variability Coefficient	Standard Deviation	Mean	Unit
	(Percent)			
Tobacco	26.38[a]	34.011	128.94	$/acre
Potatoes	24.43	11.728	48.00	$/acre
Barley	19.27	3.309	17.17	$/acre
Wheat	16.89	3.295	19.51	$/acre
Oats	12.73	1.573	12.36	$/acre
Buckwheat	11.56	1.166	10.09	$/acre
Corn	10.07	2.160	21.46	$/acre
Rye	9.36	.908	9.70	$/acre
Tame Hay	6.54	.915	13.99	$/acre

Selected Crops and Livestock, New Jersey: Ranking
By Price Random Variability Coefficients

	Random Variability Coefficient	Standard Deviation	Mean	Unit
	(Percent)			
Crop				
Potatoes	28.66	.336	1.17	$/cwt.
Sweet Potatoes	20.90[d]	.273	1.31	$/cwt.
Wheat	11.30	.129	1.15	$/bu.
Buckwheat	10.80	.080	.74	$/bu.
Tame Hay	10.78	1.751	16.25	$/ton
Oats	10.31	.044	.43	$/bu.
Corn	8.02	.049	.62	$/bu.
Rye	7.54	.057	.76	$/bu.
Livestock				
Hogs	7.33	.750	10.24	$/Head
Sheep	5.34	.233	4.37	$/Head
Horses	4.45	4.523	101.62	$/Head
Mules	3.82	4.505	117.87	$/Head
All Cattle	3.54	1.342	37.95	$/Head
Milk Cows	3.30	1.378	41.77	$/Head

Selected Crops and Livestock, New Jersey: Ranking By
Yield and Number on Farms Random Variability Coefficients

	Random Variability Coefficient	Standard Deviation	Mean	Unit
	(Percent)			
Crop				
Potatoes	25.49	14.381	56.41	Cwt./acre
Sweet Potatoes	16.54[e]	10.172	61.50	Cwt./acre
Oats	14.43	3.764	26.09	Bu./acre
Wheat	13.56	2.037	15.02	Bu./acre
Tame Hay	11.75	.150	1.28	Ton/acre
Corn	10.94	3.600	32.90	Bu./acre
Buckwheat	10.89	1.366	12.54	Bu./acre
Rye	6.29	.705	11.21	Bu./acre
Livestock				
Mules	3.04	.218	7.2	1000's of Head
Sheep	1.92	1.384	72.0	1000's of Head
All Cattle	1.10	2.570	232.7	1000's of Head
Hogs	.88	1.594	181.6	1000's of Head
Milk Cows	.46[a]	.693	151.4	1000's of Head
Horses	.31	.275	89.0	1000's of Head

TABLE A24

Selected Crops, New Jersey: Ranking By
Gross Income Random Variability Coefficients

Crop	Random Variability Coefficient	Standard Deviation	Mean	Unit
	(Percent)			
Sweet Potatoes	27.66[d]	22.196	80.25	$/acre
Potatoes	19.25	12.367	64.23	$/acre
Wheat	15.83	2.667	16.85	$/acre
Oats	14.45	1.600	11.07	$/acre
Rye	12.89	1.069	8.29	$/acre
Buckwheat	12.42	1.129	9.09	$/acre
Corn	11.32	2.291	20.23	$/acre
Tame Hay	7.64	1.559	20.41	$/acre

Selected Crops and Livestock for Pennsylvania: Ranking By Price Random Variability Coefficients

Crop	Random Variability Coefficient	Standard Deviation	Mean	Unit
	(Percent)			
Crop				
Potatoes	28.64	.287	1.00	$/cwt.
Tobacco	16.55	.018	.11	$/lb.
Oats	14.40	.058	.40	$/bu.
Wheat	12.06	.131	1.09	$/bu.
Buckwheat	11.05	.074	.67	$/bu.
Barley	10.70[b]	.078	.73	$/bu.
Corn	9.75	.058	.60	$/bu.
Rye	7.96	.058	.72	$/bu.
Tame Hay	6.73	.880	13.07	$/ton
Livestock				
Hogs	11.10	.929	8.37	$/Head
Milk Cows	5.88	1.935	32.91	$/Head
All Cattle	5.44	1.595	29.30	$/Head
Sheep	4.69	.160	3.41	$/Head
Horses	3.50	3.008	85.91	$/Head
Mules	2.85	2.887	101.32	$/Head

Selected Crops and Livestock, Pennsylvania: Ranking By
Yield and Number on Farms Random Variability Coefficients

	Random Variability Coefficient	Standard Deviation	Mean	Unit
	(Percent)			
Crop				
Tobacco	18.31	219.229	1197.50	Lb./acre
Potatoes	14.35	7.381	51.43	Cwt./acre
Wheat	14.30	2.147	15.01	Bu./acre
Oats	12.78	3.508	27.45	Bu./acre
Corn	10.24	3.499	34.17	Bu./acre
Buckwheat	8.75	1.313	15.00	Bu./acre
Rye	7.65	.861	11.26	Bu./acre
Tame Hay	7.07	.085	1.21	Ton/acre
Barley	4.03	.809	20.08	Bu./acre
Livestock				
Hogs	1.37	14.176	1032.8	1000's of Head
Mules	1.16	.333	28.6	1000's of Head
Sheep	.93	14.266	1532.3	1000's of Head
All Cattle	.59	10.168	1733.8	1000's of Head
Horses	.19	1.073	553.9	1000's of Head
Milk Cows	.15	1.275	863.8	1000's of Head

TABLE A27

Selected Crops for Pennsylvania: Ranking By
Gross Income Random Variability Coefficients

Crop	Random Variability Coefficient	Standard Deviation	Mean	Unit
	(Percent)			
Tobacco	29.14	36.983	126.91	$/acre
Potatoes	16.65	8.381	50.34	$/acre
Wheat	15.54	2.460	15.83	$/acre
Rye	12.93	1.042	8.06	$/acre
Oats	12.29	1.336	10.87	$/acre
Corn	10.94	2.213	20.23	$/acre
Buckwheat	7.99	.791	9.91	$/acre
Barley	7.61	1.096	14.40	$/acre
Tame Hay	4.78	.746	15.63	$/acre

Selected Crops and Livestock, Ohio: Ranking
By Price Random Variability Coefficients

	Random Variability Coefficient	Standard Deviation	Mean	Unit
	(Percent)			
Crop				
Potatoes	35.45	.343	.97	$/cwt.
Tobacco	22.37	.017	.07	$/lb.
Corn	18.16	.080	.44	$/bu.
Tame Hay	13.62	1.375	10.09	$/ton
Wheat	13.25	.132	1.00	$/bu.
Oats	12.99	.044	.34	$/bu.
Barley	12.21	.084	.69	$/bu.
Rye	10.85	.072	.66	$/bu.
Buckwheat	9.36	.068	.73	$/bu.
Livestock				
Hogs	9.36	.598	6.39	$/Head
Sheep	5.16	.155	3.00	$/Head
All Cattle	3.34	.940	28.12	$/Head
Mules	2.57	2.002	77.85	$/Head
Horses	2.17	1.614	74.39	$/Head
Milk Cows	1.81	.584	32.31	$/Head

Selected Crops and Livestock, Ohio: Ranking By
Yield and Number on Farms Random Variability Coefficients

	Random Variability Coefficient	Standard Deviation	Mean	Unit
	(Percent)			
Crop				
Wheat	22.82	3.445	15.10	Bu./acre
Potatoes	17.51	8.598	49.11	Cwt./acre
Rye	14.35	2.044	14.24	Bu./acre
Tobacco	14.14	124.768	882.64	Lb./acre
Buckwheat	13.34	1.789	13.42	Bu./acre
Oats	12.98	4.069	31.33	Bu./acre
Corn	12.17	4.312	35.43	Bu./acre
Tame Hay	11.99	.146	1.22	Ton/acre
Barley	10.49	2.835	27.02	Bu./acre
Livestock				
Sheep	2.10	91.505	4347.4	1000's of Head
Hogs	2.01	46.033	2287.5	1000's of Head
Mules	1.52	.279	18.4	1000's of Head
All Cattle	.83	15.545	1866.1	1000's of Head
Milk Cows	.36	2.781	763.9	1000's of Head
Horses	.35	2.836	818.3	1000's of Head

TABLE A30

Selected Crops, Ohio: Ranking By
Gross Income Random Variability Coefficients

Crop	Random Variability Coefficient	Standard Deviation	Mean	Unit
	(Percent)			
Tobacco	28.72[a]	18.940	65.94	$/acre
Potatoes	27.30	12.504	45.80	$/acre
Wheat	16.58	2.430	14.65	$/acre
Rye	16.54	1.552	9.38	$/acre
Buckwheat	13.60	1.322	9.72	$/acre
Barley	12.94	2.371	18.32	$/acre
Corn	11.44	1.762	15.41	$/acre
Oats	10.35	1.078	10.42	$/acre
Tame Hay	8.11	.976	12.04	$/acre

Selected Crops and Livestock, Indiana: Ranking
By Price Random Variability Coefficients

	Random Variability Coefficient	Standard Deviation	Mean	Unit
	(Percent)			
Crop				
Potatoes	37.74	.359	.95	$/cwt.
Tobacco	25.87	.018	.07	$/lb.
Corn	20.21	.079	.39	$/bu.
Tame Hay	15.82	1.479	9.35	$/ton
Sweet Potatoes	15.36[d]	.232	1.51	$/cwt.
Oats	14.54	.045	.31	$/bu.
Wheat	13.88	.131	.95	$/bu.
Buckwheat	12.13	.085	.70	$/bu.
Rye	11.35	.072	.63	$/bu.
Barley	8.32	.057	.68	$/bu.
Livestock				
Hogs	10.47	.624	5.96	$/Head
Sheep	5.18	.156	3.01	$/Head
Mules	4.12	3.053	74.14	$/Head
All Cattle	4.01	1.007	25.12	$/Head
Milk Cows	3.43	1.023	29.81	$/Head
Horses	3.41	2.362	69.21	$/Head

Selected Crops and Livestock, Indiana: Ranking By
Yield and Number on Farms Random Variability Coefficients

	Random Variability Coefficient	Standard Deviation	Mean	Unit
	(Percent)			
Crop				
Potatoes	25.54	10.041	39.32	Cwt./acre
Wheat	23.47	3.301	14.07	Bu./acre
Corn	16.47	5.462	33.17	Bu./acre
Sweet Potatoes	16.00[e]	7.223	45.14	Cwt./acre
Buckwheat	15.21	1.646	10.82	Bu./acre
Oats	15.21	4.286	28.18	Bu./acre
Tobacco	14.12	115.370	817.23	Lb./acre
Rye	12.96	1.626	12.55	Bu./acre
Tame Hay	11.65	.142	1.22	Ton/acre
Barley	10.05	2.400	23.90	Bu./acre
Livestock				
Hogs	3.99	120.020	3010.4	1000's of Head
Sheep	2.80	34.222	1220.8	1000's of Head
All Cattle	1.08	15.707	1451.8	1000's of Head
Mules	.75	.472	62.7	1000's of Head
Horses	.51	3.428	676.5	1000's of Head
Milk Cows	.36	1.846	517.4	1000's of Head

TABLE A33

Selected Crops, Indiana: Ranking By
Gross Income Random Variability Coefficients

Crop	Random Variability Coefficient	Standard Deviation	Mean	Unit
	(Percent)			
Tobacco	21.08	12.171	57.73	$/acre
Wheat	20.06	2.673	12.97	$/acre
Rye	17.15	1.361	7.94	$/acre
Potatoes	16.28	5.670	34.84	$/acre
Sweet Potatoes	12.71[a,d]	8.814	69.37	$/acre
Corn	13.07	1.654	12.65	$/acre
Oats	12.04	1.042	8.66	$/acre
Buckwheat	11.63	.878	7.55	$/acre
Barley	9.19	1.478	16.08	$/acre
Tame Hay	8.60	.969	11.27	$/acre

Selected Crops and Livestock, Illinois: Ranking
By Price Random Variability Coefficients

	Random Variability Coefficient	Standard Deviation	Mean	Unit
	(Percent)			
Crop				
Potatoes	39.35	.394	1.00	$/cwt.
Corn	30.25	.112	.37	$/bu.
Oats	28.08	.082	.29	$/bu.
Tobacco	22.88	.018	.08	$/lb.
Tame Hay	19.86	1.697	8.54	$/ton
Sweet Potatoes	16.09[d]	.245	1.52	$/cwt.
Wheat	15.03	.133	.88	$/bu.
Barley	13.86	.083	.60	$/bu.
Buckwheat	12.31	.090	.73	$/bu.
Rye	12.19	.071	.58	$/bu.
Livestock				
Hogs	10.43	.680	6.52	$/Head
Sheep	5.94	.184	3.10	$/Head
All Cattle	4.20	1.101	26.19	$/Head
Milk Cows	3.73	1.159	31.07	$/Head
Horses	3.64	2.485	68.34	$/Head
Mules	3.41	2.599	76.21	$/Head

TABLE A35

Selected Crops and Livestock, Illinois: Ranking By
Yield and Number on Farms Random Variability Coefficients

	Random Variability Coefficient	Standard Deviation	Mean	Unit
	(Percent)			
Crop				
Potatoes	27.76	12.787	46.07	Cwt./acre
Wheat	21.64	2.955	13.66	Bu./acre
Corn	19.61.	6.399	32.62	Bu./acre
Sweet Potatoes	18.49[e]	7.840	42.41	Cwt./acre
Buckwheat	17.17	1.866	10.86	Bu./acre
Oats	16.43	5.217	31.76	Bu./acre
Tame Hay	15.21	.198	1.30	Ton/acre
Tobacco	9.36	63.359	677.16	Lb./acre
Barley	8.73	2.222	25.47	Bu./acre
Rye	4.60	.718	15.60	Bu./acre
Livestock				
Hogs	3.46	155.187	4487.5	1000's of Head
Sheep	2.56	25.108	982.6	1000's of Head
All Cattle	1.61	39.275	2444.9	1000's of Head
Mules	.99	1.167	117.4	1000's of Head
Milk Cows	.75	6.382	853.7	1000's of Head
Horses	.35	4.221	1204.0	1000's of Head

TABLE A36

Selected Crops, Illinois: Ranking By
Gross Income Random Variability Coefficients

Crop	Random Variability Coefficient	Standard Deviation	Mean	Unit
	(Percent)			
Tobacco	25.98	13.648	52.53	$/acre
Wheat	20.50	2.417	11.79	$/acre
Oats	16.22	1.477	9.11	$/acre
Corn	15.11	1.770	11.72	$/acre
Barley	14.82	2.211	14.92	$/acre
Sweet Potatoes	14.19[d]	8.663	61.06	$/acre
Buckwheat	13.42	1.067	7.95	$/acre
Rye	11.94	1.088	9.12	$/acre
Tame Hay	11.11	1.215	10.94	$/acre
Potatoes	10.94	4.631	42.35	$/acre

Selected Crops and Livestock, Michigan: Ranking
By Price Random Variability Coefficients

	Random Variability Coefficient	Standard Deviation	Mean	Unit
	(Percent)			
Crop				
Potatoes	36.94	.284	.77	$/cwt.
Oats	16.67	.060	.36	$/bu.
Tame Hay	15.49	1.662	10.73	$/ton
Wheat	13.46	.133	.99	$/bu.
Rye	12.46	.080	.64	$/bu.
Buckwheat	12.13	.075	.62	$/bu.
Corn	10.18	.051	.50	$/bu.
Barley	8.37	.057	.68	$/bu.
Livestock				
Hogs	7.82	.504	6.44	$/Head
Sheep	6.22	.187	3.01	$/Head
Mules	5.76	4.956	86.01	$/Head
Horses	4.45	3.527	79.36	$/Head
All Cattle	4.05	1.120	27.65	$/Head
Milk Cows	2.62	.851	32.43	$/Head

TABLE A38

Selected Crops and Livestock, Michigan: Ranking By
Yield and Number on Farms Random Variability Coefficients

	Random Variability Coefficient	Standard Deviation	Mean	Unit
	(Percent)			
Crop				
Potatoes	18.98	9.681	51.00	Cwt./acre
Wheat	16.34	2.652	16.23	Bu./acre
Buckwheat	12.11	1.540	12.72	Bu./acre
Corn	11.24	3.639	32.38	Bu./acre
Oats	11.06	3.561	32.20	Bu./acre
Tame Hay	10.30	.126	1.22	Ton/acre
Rye	8.87	1.209	13.63	Bu./acre
Barley	8.08	1.888	23.36	Bu./acre
Livestock				
Mules	5.56	.194	3.5	1000's of Head
All Cattle	2.33	23.348	1003.4	1000's of Head
Hogs	1.61	13.900	864.8	1000's of Head
Sheep	1.46	30.994	2116.1	1000's of Head
Milk Cows	.46	2.109	463.2	1000's of Head
Horses	.34	1.564	462.4	1000's of Head

Selected Crops, Michigan: Ranking By
Gross Income Random Variability Coefficients

Crop	Random Variability Coefficient	Standard Deviation	Mean	Unit
	(Percent)			
Potatoes	22.97	8.769	38.18	$/acre
Oats	14.03	1.609	11.47	$/acre
Wheat	13.96	2.215	15.87	$/acre
Tame Hay	12.91	1.682	13.03	$/acre
Rye	11.50	1.018	8.85	$/acre
Buckwheat	11.04	.872	7.90	$/acre
Corn	9.58	1.532	16.00	$/acre
Barley	7.76	1.240	16.00	$/acre

TABLE A40

Selected Crops and Livestock, Wisconsin: Ranking
By Price Random Variability Coefficients

	Random Variability Coefficient	Standard Deviation	Mean	Unit
	(Percent)			
Crop				
Potatoes	39.00	.296	.76	$/cwt.
Tame Hay	24.42	2.137	8.75	$/ton
Rye	17.39	.102	.59	$/bu.
Wheat	16.98	.143	.84	$/bu.
Oats	16.88	.054	.32	$/bu.
Tobacco	15.92[a]	.016	.10	$/lb.
Barley	15.18	.091	.60	$/bu.
Buckwheat	14.95	.093	.62	$/bu.
Corn	9.46	.042	.44	$/bu
Livestock				
Hogs	13.20	.883	6.69	$/Head
Mules	6.24	5.197	83.23	$/Head
Sheep	5.32	.148	2.78	$/Head
All Cattle	4.16	1.006	24.15	$/Head
Milk Cows	4.09	1.161	28.39	$/Head
Horses	3.33	2.525	75.73	$/Head

Selected Crops and Livestock, Wisconsin: Ranking By
Yield and Number on Farms Random Variability Coefficients

	Random Variability Coefficient	Standard Deviation	Mean	Unit
	(Percent)			
Crop				
Buckwheat	23.00	2.561	11.14	Bu./acre
Potatoes	18.69	10.544	56.41	Cwt./acre
Tame Hay	13.71	.181	1.32	Ton/acre
Corn	13.50	4.394	32.55	Bu./acre
Wheat	11.97	1.738	14.52	Bu./acre
Barley	11.35	3.073	27.08	Bu./acre
Tobacco	10.88	119.657	1100.24	Lb./acre
Oats	10.82	3.567	32.98	Bu./acre
Rye	7.28	1.037	14.24	Bu./acre
Livestock				
Mules	5.23	.252	4.8	1000's of Head
Hogs	4.83	50.833	1051.7	1000's of Head
Sheep	1.64	18.489	1129.7	1000's of Head
All Cattle	1.32	20.275	1535.4	1000's of Head
Milk Cows	.50	3.575	716.0	1000's of Head
Horses	.33	1.468	441.0	1000's of Head

TABLE A42

Selected Crops, Wisconsin: Ranking By
Gross Income Random Variability Coefficients

Crop	Random Variability Coefficient	Standard Deviation	Mean	Unit
	(Percent)			
Potatoes	34.96	14.423	41.26	$/acre
Tame Hay	23.24	2.659	11.44	$/acre
Tobacco	21.63[a]	22.203	102.66	$/acre
Buckwheat	18.48	1.254	6.78	$/acre
Wheat	18.45	2.217	12.01	$/acre
Corn	17.62	2.515	14.28	$/acre
Barley	17.57	2.778	15.81	$/acre
Rye	17.23	1.433	8.32	$/acre
Oats	14.97	1.566	10.46	$/acre

Selected Crops and Livestock, Minnesota: Ranking
By Price Random Variability Coefficients[1]

	Random Variability Coefficient	Standard Deviation	Mean	Unit
	(Percent)			
Crop				
Potatoes	43.22	.304	.70	$/cwt.
Barley	18.66	.091	.49	$/bu.
Wheat	16.27	.123	.75	$/bu.
Oats	16.19	.049	.30	$/bu.
Rye	13.59	.070	.52	$/bu.
Tame Hay	13.11	.717	5.47	$/ton
Corn	11.33	.045	.40	$/bu.
Buckwheat	10.61	.068	.64	$/bu.
Livestock				
Mules	10.69[a]	8.771	82.07	$/Head
Hogs	9.50	.599	6.30	$/Head
Sheep	4.98	.133	2.67	$/Head
All Cattle	4.45	1.002	22.51	$/Head
Milk Cows	3.06	.806	26.31	$/Head
Horses	3.04	2.222	73.09	$/Head

[1] All estimates are based on 43 observations, 1867-1909.

Selected Crops and Livestock, Minnesota: Ranking By
Yield and Number on Farms Random Variability Coefficients[1]

	Random Variability Coefficient	Standard Deviation	Mean	Unit
	(Percent)			
Crop				
Potatoes	16.78	9.765	58.19	Cwt./acre
Wheat	15.82	2.266	14.27	Bu./acre
Buckwheat	15.73	1.857	11.80	Bu./acre
Corn	13.55	3.990	29.45	Bu./acre
Oats	13.50	4.400	32.59	Bu./acre
Tame Hay	11.87	.173	1.45	Ton/acre
Barley	7.86	1.966	25.01	Bu./acre
Rye	5.34	.892	16.69	Bu./acre
Livestock				
Mules	4.03	.292	7.3	1000's of Head
Hogs	3.48	26.831	770.9	1000's of Head
Sheep	3.01	9.077	301.3	1000's of Head
All Cattle	1.36	15.974	1172.8	1000's of Head
Milk Cows	1.24	6.799	547.1	1000's of Head
Horses	1.09	4.667	426.5	1000's of Head

[1]All estimates are based on 43 observations, 1867–1909.

TABLE A45

Selected Crops, Minnesota: Ranking By
Gross Income Random Variability Coefficients[1]

Crop	Random Variability Coefficient	Standard Deviation	Mean	Unit
	(Percent)			
Wheat	28.62	3.065	10.71	$/acre
Potatoes	25.68	10.083	39.26	$/acre
Oats	17.04	1.666	9.78	$/acre
Rye	14.73	1.267	8.60	$/acre
Buckwheat	13.62	1.024	7.52	$/acre
Barley	12.29	1.502	12.22	$/acre
Corn	10.43	1.232	11.81	$/acre
Tame Hay	9.85	.785	7.97	$/acre

[1]All estimates are based on 43 observations, 1867–1909.

Selected Crops and Livestock, Iowa: Ranking
By Price Random Variability Coefficients

	Random Variability Coefficient	Standard Deviation	Mean	Unit
	(Percent)			
Crop				
Potatoes	46.50	.377	.81	$/cwt.
Corn	33.03	.106	.32	$/bu.
Barley	21.87	.104	.48	$/bu.
Oats	21.08	.055	.26	$/bu.
Rye	20.57	.104	.51	$/bu.
Wheat	18.39	.137	.74	$/bu.
Sweet Potatoes	17.06[d]	.303	1.78	$/cwt.
Buckwheat	11.60	.082	.71	$/bu.
Tame Hay	10.13	.585	5.78	$/ton
Livestock				
Hogs	9.87[b]	.658	6.66	$/Head
All Cattle	5.03	1.225	24.34	$/Head
Milk Cows	4.74	1.339	28.23	$/Head
Mules	4.22	3.257	77.14	$/Head
Sheep	4.03	.121	3.01	$/Head
Horses	3.38	2.277	67.44	$/Head

Selected Crops and Livestock, Iowa: Ranking By
Yield and Number on Farms Random Variability Coefficients

	Random Variability Coefficient	Standard Deviation	Mean	Unit
	(Percent)			
Crop				
Potatoes	22.54	11.376	50.48	Cwt./acre
Sweet Potatoes	16.64[1]	7.804	46.89	Cwt./acre
Corn	15.96	5.765	36.13	Bu./acre
Oats	14.99	4.896	32.66	Bu./acre
Wheat	14.08	1.852	13.15	Bu./acre
Buckwheat	12.84	1.429	11.13	Bu./acre
Tame Hay	11.98	.172	1.43	Ton/acre
Barley	9.09	2.157	23.73	Bu./acre
Rye	6.89	1.030	14.96	Bu./acre
Livestock				
Sheep	3.86	25.749	667.3	1000's of Head
Hogs	3.05	158.519	5203.5	1000's of Head
All Cattle	2.16	72.000	3338.3	1000's of Head
Mules	1.96	.889	45.3	1000's of Head
Horses	1.02	11.099	1085.7	1000's of Head
Milk Cows	.94	9.458	1010.8	1000's of Head

[1]This estimate is based on 36 observations, 1874-1909.

TABLE A48

Selected Crops, Iowa: Ranking By
Gross Income Random Variability Coefficients

Crop	Random Variability Coefficient	Standard Deviation	Mean	Unit
	(Percent)			
Potatoes	35.34	13.356	37.80	$/acre
Sweet Potatoes	23.60[d]	19.370	82.08	$/acre
Wheat	19.93	1.917	9.62	$/acre
Tame Hay	19.10	1.565	8.20	$/acre
Rye	17.59	1.339	7.61	$/acre
Barley	16.66	1.835	11.01	$/acre
Oats	14.14	1.184	8.38	$/acre
Corn	9.55	1.079	11.30	$/acre
Buckwheat	6.60	.517	7.84	$/acre

Selected Crops and Livestock, Missouri: Ranking
By Price Random Variability Coefficients

	Random Variability Coefficient	Standard Deviation	Mean	Unit
	(Percent)			
Crop				
Corn	39.93	.138	.38	$/bu.
Potatoes	38.91	.364	.94	$/cwt.
Tame Hay	24.59	2.025	8.24	$/ton
Oats	19.48	.059	.30	$/bu.
Sweet Potatoes	19.29[d]	.248	1.29	$/cwt.
Wheat	15.79	.137	.87	$/bu.
Rye	14.83	.090	.61	$/bu.
Barley	14.21	.092	.65	$/bu.
Buckwheat	12.42	.089	.72	$/bu.
Tobacco	12.22	.011	.09	$/lb.
Livestock				
Hogs	16.35	.730	4.47	$/Head
Mules	7.86	5.249	66.77	$/Head
Sheep	5.72	.136	2.38	$/Head
All Cattle	5.38	1.120	20.80	$/Head
Milk Cows	4.81	1.172	24.38	$/Head
Horses	4.25[b]	2.326	54.73	$/Head

Selected Crops and Livestock, Missouri: Ranking By
Yield and Number on Farms Random Variability Coefficients

	Random Variability Coefficient	Standard Deviation	Mean	Unit
	(Percent)			
Crop				
Potatoes	24.88	10.964	44.07	Cwt./acre
Oats	20.78	4.769	22.95	Bu./acre
Buckwheat	19.83	2.133	10.76	Bu./acre
Sweet Potatoes	19.71[e]	8.423	42.74	Cwt./acre
Corn	19.20	5.542	28.86	Bu./acre
Wheat	17.26	2.268	13.14	Bu./acre
Tame Hay	15.69	.181	1.15	Ton/acre
Tobacco	15.62	128.293	821.14	Lb./acre
Rye	13.41	1.478	11.02	Bu./acre
Barley	9.58	1.845	19.25	Bu./acre
Livestock				
Hogs	2.86	109.352	3820.1	1000's of Head
Sheep	1.90[b]	19.878	1043.0	1000's of Head
All Cattle	1.79	41.854	2334.2	1000's of Head
Milk Cows	1.52	9.605	630.2	1000's of Head
Mules	1.11[b]	2.606	234.4	1000's of Head
Horses	.81	6.796	840.3	1000's of Head

TABLE A51

Selected Crops, Missouri: Ranking By
Gross Income Random Variability Coefficients

Crop	Random Variability Coefficient	Standard Deviation	Mean	Unit
	(Percent)			
Buckwheat	20.80	1.615	7.76	$/acre
Potatoes	19.29	7.421	38.47	$/acre
Sweet Potatoes	18.92[d]	11.078	58.56	$/acre
Rye	16.78	1.123	6.69	$/acre
Barley	16.72	2.076	12.41	$/acre
Corn	15.97	1.699	10.64	$/acre
Wheat	15.84	1.789	11.29	$/acre
Tobacco	14.36	10.364	72.19	$/acre
Cotton	13.47[1]	2.943	21.85	$/acre
Oats	13.39	.931	6.95	$/acre
Tame Hay	11.17	1.050	9.40	$/acre

[1]This estimate is based on 15 observations, 1895-1909.

Selected Crops and Livestock, North Dakota: Ranking
By Price Random Variability Coefficients[1]

	Random Variability Coefficient	Standard Deviation	Mean	Unit
	(Percent)			
Crop				
Potatoes	30.79	.207	.67	$/cwt.
Rye	26.41[a]	.121	.46	$/bu.
Barley	24.58	.091	.37	$/bu.
Wheat	24.43	.155	.64	$/bu.
Tame Hay	17.59	.730	4.15	$/ton
Corn	14.67[2]	.059	.40	$/bu.
Buckwheat	14.38	.095	.66	$/bu.
Oats	11.72	.033	.28	$/bu.
Livestock				
Hogs	9.82	.678	6.91	$/Head
Milk Cows	8.92	2.381	26.69	$/Head
Sheep	5.99	.169	2.82	$/Head
All Cattle	5.81	1.247	21.46	$/Head
Mules	4.13	3.524	85.26	$/Head
Horses	2.92[b]	2.015	69.11	$/Head

[1]All estimates are based on 28 observations unless otherwise noted, 1882–1909.

[2]This estimate is based on 14 observations, 1882–1895.

Selected Crops and Livestock, North Dakota: Ranking By
Yield and Number on Farms Random Variability Coefficients[1]

	Random Variability Coefficient	Standard Deviation	Mean	Unit
	(Percent)			
Crop				
Wheat	27.35	3.912	14.30	Bu./acre
Potatoes	26.16	13.416	51.29	Cwt./acre
Rye	23.98	2.748	11.46	Bu./acre
Oats	22.94	6.304	27.48	Bu./acre
Corn	22.07	4.695	21.28	Bu./acre
Barley	19.16	4.107	21.43	Bu./acre
Tame Hay	13.98	.174	1.25	Ton/acre
Livestock				
Sheep	7.70	25.085	325.6	1000's of Head
Hogs	6.30[b]	8.290	131.6	1000's of Head
All Cattle	4.72	23.034	488.2	1000's of Head
Mules	4.47	.330	7.4	1000's of Head
Horses	1.73	5.055	292.4	1000's of Head
Milk Cows	1.44	1.499	103.9	1000's of Head

[1]All estimates are based on 28 observations, 1882-1909.

TABLE A54

Selected Crops, North Dakota: Ranking By
Gross Income Random Variability Coefficients

Crop	Random Variability Coefficient	Standard Deviation	Mean	Unit
	(Percent)			
Rye	37.94	2.018	5.32	$/acre
Wheat	34.90	3.172	9.09	$/acre
Potatoes	33.46	10.999	32.88	$/acre
Oats	27.65	2.116	7.65	$/acre
Barley	25.12	1.997	7.95	$/acre
Tame Hay	17.76[a]	.908	5.11	$/acre
Corn	16.50	1.397	8.47	$/acre

[1]All estimates are based on 28 observations, 1882-1909.

Selected Crops and Livestock, South Dakota: Ranking
By Price Random Variability Coefficients[1]

	Random Variability Coefficient	Standard Deviation	Mean	Unit
	(Percent)			
Crop				
Potatoes	37.59	.271	.72	$/cwt.
Corn	26.64	.097	.35	$/bu.
Barley	27.26	.101	.37	$/bu.
Rye	25.05[a]	.114	.46	$/bu.
Oats	24.58	.066	.27	$/bu.
Wheat	24.09	.153	.63	$/bu.
Buckwheat	14.02[2]	.091	.65	$/bu.
Tame Hay	6.74	.271	4.02	$/ton
Livestock				
Hogs	11.05	.755	6.83	$/Head
Milk Cows	9.14	2.378	26.02	$/Head
All Cattle	5.21	1.123	21.54	$/Head
Sheep	5.02	.144	2.88	$/Head
Mules	3.35	2.505	74.73	$/Head
Horses	2.15	1.327	61.63	$/Head

[1]All estimates are based on 28 observations unless otherwise noted, 1882-1909.

[2]This estimate is based on 14 observations, 1882-1896.

Selected Crops and Livestock, South Dakota: Ranking By [1]
Yield and Number on Farms Random Variability Coefficients

	Random Variability Coefficient	Standard Deviation	Mean	Unit
	(Percent)			
Crop				
Potatoes	32.74	13.330	40.71	Cwt./acre
Barley	30.07	6.100	20.28	Bu./acre
Corn	26.54	6.360	23.96	Bu./acre
Oats	25.95	6.947	26.77	Bu./acre
Wheat	25.14	2.926	11.64	Bu./acre
Rye	16.42	1.875	11.42	Bu./acre
Tame Hay	12.79	.179	1.40	Ton/acre
Livestock				
Hogs	6.50[a]	30.022	492.1	1000's of Head
Sheep	4.16	16.827	404.9	1000's of Head
Mules	4.00	.277	6.9	1000's of Head
All Cattle	2.64	27.512	1044.1	1000's of Head
Milk Cows	1.62	3.595	221.4	1000's of Head
Horses	.76	2.841	374.4	1000's of Head

[1]All estimates are based on 28 observations, 1882-1909.

TABLE A57

Selected Crops, South Dakota: Ranking By
Gross Income Random Variability Coefficients[1]

Crop	Random Variability Coefficient	Standard Deviation	Mean	Unit
	(Percent)			
Barley	40.11	3.009	7.50	$/acre
Potatoes	24.35	6.787	27.87	$/acre
Wheat	22.77	1.699	7.46	$/acre
Rye	22.53	1.202	5.34	$/acre
Oats	13.05	.932	7.15	$/acre
Corn	12.54	1.025	8.18	$/acre
Tame Hay	9.20	.519	5.65	$/acre

Selected Crops and Livestock, Nebraska: Ranking
By Price Random Variability Coefficients

	Random Variability Coefficient	Standard Deviation	Mean	Unit
	(Percent)			
Crop				
Potatoes	59.96	.558	.93	$/cwt.
Corn	48.83	.157	.32	$/bu.
Oats	27.54	.074	.27	$/bu.
Tame Hay	24.68	1.061	4.30	$/ton
Rye	21.78[a]	.103	.48	$/bu.
Wheat	20.38	.138	.68	$/bu.
Barley	17.80	.079	.45	$/bu.
Buckwheat	10.90[1]	.081	.74	$/bu.
Livestock				
Hogs	14.38	.917	6.37	$/Head
Mules	7.21[b]	6.106	84.37	$/Head
Sheep	5.23	.142	2.72	$/Head
All Cattle	5.09	1.178	23.15	$/Head
Horses	5.00	3.319	66.42	$/Head
Milk Cows	4.92	1.404	28.54	$/Head

[1]This estimate is based on 43 observations, 1867-1909.

TABLE A59

Selected Crops and Livestock, Nebraska: Ranking By
Yield and Number on Farms Random Variability Coefficients

	Random Variability Coefficient	Standard Deviation	Mean	Unit
	(Percent)			
Crop				
Corn	25.98	7.935	30.55	Bu./acre
Potatoes	24.70	10.677	43.23	Cwt./acre
Barley	23.12	4.410	19.08	Bu./acre
Oats	19.95[1]	5.549	27.81	Bu./acre
Buckwheat	16.86[1]	1.425	8.45	Bu./acre
Tame Hay	15.76	.277	1.76	Ton/acre
Rye	13.17	1.696	12.88	Bu./acre
Wheat	10.68	1.423	13.32	Bu./acre

	Random Variability Coefficient	Standard Deviation	Mean	Unit
	(Percent)			
Livestock				
Hogs	10.52[b]	206.526	1963.3	1000's of Head
Sheep	6.31	15.634	247.6	1000's of Head
All Cattle	2.86	51.148	1789.3	1000's of Head
Horses	1.98	9.890	498.7	1000's of Head
Mules	1.62	.607	37.5	1000's of Head
Milk Cows	1.39	4.208	302.3	1000's of Head

[1]This estimate is based on 43 observations, 1867–1909.

TABLE A60

Selected Crops, Nebraska: Ranking By
Gross Income Random Variability Coefficients

Crop	Random Variability Coefficient	Standard Deviation	Mean	Unit
	(Percent)			
Potatoes	26.39	9.723	36.84	$/acre
Rye	22.94	1.413	6.16	$/acre
Barley	22.25[b]	1.876	8.43	$/acre
Wheat	18.38	1.643	8.94	$/acre
Oats	17.86	1.321	7.39	$/acre
Buckwheat	15.04[1]	.968	6.44	$/acre
Corn	14.80	1.376	9.30	$/acre
Tame Hay	14.10	1.047	7.43	$/acre

[1]This estimate is based on 43 observations, 1867–1909.

Selected Crops and Livestock, Kansas: Ranking
By Price Random Variability Coefficients

	Random Variability Coefficient	Standard Deviation	Mean	Unit
	(Percent)			
Crop				
Corn	49.08	.182	.37	$/bu.
Potatoes	40.58	.458	1.13	$/cwt.
Tame Hay	27.84	1.311	4.71	$/ton
Oats	25.05	.075	.30	$/bu.
Sweet Potatoes	19.93[d]	.302	1.51	$/cwt.
Wheat	19.75	.156	.79	$/bu.
Buckwheat	17.77[1]	.133	.75	$/bu.
Rye	17.54[2]	.087	.50	$/bu.
Barley	14.67	.072	.49	$/bu.
Livestock				
Hogs	16.63	1.042	6.27	$/Head
Sheep	7.30	.188	2.58	$/Head
All Cattle	5.39	1.215	22.54	$/Head
Milk Cows	4.26	1.138	26.72	$/Head
Mules	3.28[b]	2.348	71.54	$/Head
Horses	2.75	1.604	58.42	$/Head

[1] This estimate is based on 18 observations, 1878-1895.

[2] This estimate is based on 41 observations, 1869-1909.

Selected Crops and Livestock, Kansas: Ranking By
Yield and Number on Farms Random Variability Coefficients

	Random Variability Coefficient	Standard Deviation	Mean	Unit
	(Percent)			
Crop				
Barley	35.77	6.125	17.12	Bu./acre
Potatoes	34.27	12.969	37.84	Cwt./acre
Corn	31.07	8.080	26.01	Bu./acre
Wheat	22.30	3.124	14.01	Bu./acre
Tame Hay	20.09	.368	1.83	Ton/acre
Sweet Potatoes	19.70[e]	7.664	38.91	Cwt./acre
Oats	16.78	4.585	27.33	Bu./acre
Rye	13.94[1]	1.710	12.26	Bu./acre
Livestock				
Hogs	5.23	102.253	1956.5	1000's of Head
Sheep	4.79	20.582	429.9	1000's of Head
Horses	1.64	11.479	700.2	1000's of Head
All Cattle	1.43	31.625	2216.5	1000's of Head
Mules	1.38	1.189	86.4	1000's of Head
Milk Cows	.87	4.200	485.4	1000's of Head

[1] This estimate is based on 41 observations, 1869-1909.

TABLE A63

Selected Crops, Kansas: Ranking By
Gross Income Random Variability Coefficients

Crop	Random Variability Coefficient	Standard Deviation	Mean	Unit
	(Percent)			
Barley	26.42	2.268	8.59	$/acre
Potatoes	23.43	9.433	40.27	$/acre
Rye	18.84[1]	1.147	6.09	$/acre
Wheat	15.32	1.704	11.12	$/acre
Oats	15.01	1.207	8.04	$/acre
Sweet Potatoes	13.57[d]	9.008	66.39	$/acre
Corn	13.15	1.172	8.91	$/acre
Tame Hay	7.06	.593	8.40	$/acre

[1] This estimate is based on 41 observations, 1869-1909.

Selected Crops and Livestock, Delaware: Ranking
By Price Random Variability Coefficients

	Random Variability Coefficient	Standard Deviation	Mean	Unit
	(Percent)			
Crop				
Potatoes	27.13	.291	1.07	$/cwt.
Sweet Potatoes	14.76[d]	.146	.99	$/cwt.
Tame Hay	14.25	2.171	15.23	$/ton
Buckwheat	13.93[1]	.080	.58	$/bu.
Wheat	12.55	.138	1.10	$/bu.
Oats	10.05	.040	.40	$/bu.
Corn	9.80	.050	.51	$/bu.
Rye	9.19[2]	.073	.79	$/bu.
Livestock				
Hogs	20.33	1.457	7.16	$/Head
All Cattle	9.08	2.519	27.75	$/Head
Milk Cows	8.50	2.673	31.46	$/Head
Mules	8.44	8.567	101.55	$/Head
Sheep	6.86	.249	3.63	$/Head
Horses	6.24	5.052	80.94	$/Head

[1]This estimate is based on 29 observations, 1881-1909.

[2]This estimate is based on 27 observations, 1866-1892.

Selected Crops and Livestock, Delaware: Ranking By
Yield and Number on Farms Random Variability Coefficients

	Random Variability Coefficient	Standard Deviation	Mean	Unit
	(Percent)			
Crop				
Rye	22.07[1]	1.924	8.72	Bu./acre
Wheat	18.41	2.700	14.67	Bu./acre
Potatoes	16.43	8.290	50.46	Cwt./acre
Tame Hay	13.41	.151	1.13	Ton/acre
Buckwheat	11.18[2]	1.601	14.32	Bu./acre
Sweet Potatoes	11.02[e]	5.989	54.33	Cwt./acre
Oats	9.50[a]	2.125	22.37	Bu./acre
Corn	8.01[a]	1.726	21.54	Bu./acre
Livestock				
Sheep	3.51	.545	15.5	1000's of Head
Mules	3.18[b]	.144	4.5	1000's of Head
Horses	1.27	.321	25.3	1000's of Head
Hogs	1.01	.427	42.1	1000's of Head
Milk Cows	1.01	.289	28.6	1000's of Head
All Cattle	.76	.428	56.2	1000's of Head

[1]This estimate is based on 27 observations, 1866-1892.

[2]This estimate is based on 14 observations, 1896-1909.

TABLE A66

Selected Crops, Delaware: Ranking By
Gross Income Random Variability Coefficients

Crop	Random Variability Coefficient	Standard Deviation	Mean	Unit
	(Percent)			
Rye	29.08[1]	1.956	6.73	$/acre
Buckwheat	28.41[2]	2.230	7.85	$/acre
Potatoes	26.03[a]	13.901	53.41	$/acre
Wheat	19.35	2.984	15.43	$/acre
Sweet Potatoes	14.50[d]	8.155	56.26	$/acre
Oats	11.30	.998	8.23	$/acre
Corn	11.22	1.214	10.82	$/acre
Tame Hay	10.89	1.836	16.86	$/acre

[1]This estimate is based on 27 observations, 1866-1892.
[2]This estimate is based on 14 observations, 1896-1909.

Selected Crops and Livestock, Maryland: Ranking
By Price Random Variability Coefficients

	Random Variability Coefficient	Standard Deviation	Mean	Unit
	(Percent)			
Crop				
Potatoes	26.33	.277	1.05	$/cwt.
Sweet Potatoes	14.59[d]	.151	1.04	$/cwt.
Tobacco	13.59	.010	.07	$/lb.
Wheat	12.79	.140	1.10	$/bu.
Tame Hay	11.58	1.668	14.40	$/ton
Oats	10.16	.040	.40	$/bu.
Buckwheat	9.98	.070	.71	$/bu.
Rye	9.37	.067	.71	$/bu.
Corn	7.89	.042	.54	$/bu.
Livestock				
Hogs	9.93	.662	6.67	$/Head
Mules	6.71	6.770	100.87	$/Head
Sheep	5.03	.185	3.67	$/Head
Horses	3.92	3.017	77.03	$/Head
All Cattle	3.60	.943	26.19	$/Head
Milk Cows	3.28	.997	30.43	$/Head

Selected Crops and Livestock, Maryland: Ranking By
Yield and Number on Farms Random Variability Coefficients

	Random Variability Coefficient	Standard Deviation	Mean	Unit
	(Percent)			
Crop				
Oats	19.26	4.155	21.58	Bu./acre
Barley	15.55[1]	4.268	27.44	Bu./acre
Wheat	13.13	2.061	15.70	Bu./acre
Buckwheat	13.13	1.894	14.43	Bu./acre
Tobacco	12.46	80.570	646.59	Lb./acre
Rye	11.43	1.427	12.49	Bu./acre
Tame Hay	11.30	.123	1.09	Ton/acre
Sweet Potatoes	10.86[e]	5.598	51.55	Cwt./acre
Potatoes	8.89	4.192	47.14	Cwt./acre
Corn	8.21	2.117	25.79	Bu./acre
Livestock				
Hogs	2.39	6.731	282.0	1000's of Head
Mules	2.11	.302	14.3	1000's of Head
Sheep	1.24	1.721	139.3	1000's of Head
All Cattle	.76	2.085	276.0	1000's of Head
Milk Cows	.46	.602	130.6	1000's of Head
Horses	.43	.540	126.6	1000's of Head

[1] This estimate is based on 21 observations, 1889-1909.

TABLE A69

Selected Crops, Maryland: Ranking By
Gross Income Random Variability Coefficients

Crop	Random Variability Coefficient	Standard Deviation	Mean	Unit
	(Percent)			
Wheat	18.65	3.047	16.34	$/acre
Potatoes	17.40	8.549	49.13	$/acre
Tobacco	15.17	6.929	45.66	$/acre
Oats	13.81	1.159	8.40	$/acre
Sweet Potatoes	13.37[d]	7.615	56.97	$/acre
Buckwheat	13.14	1.315	10.01	$/acre
Corn	12.94	1.770	13.68	$/acre
Rye	12.65	1.115	8.82	$/acre
Tame Hay	12.37	1.898	15.34	$/acre
Barley	12.24[1]	1.640	13.40	$/acre

[1] This estimate is based on 21 observations, 1889-1909.

Selected Crops and Livestock, Virginia: Ranking
By Price Random Variability Coefficients

	Random Variability Coefficient	Standard Deviation	Mean	Unit
	(Percent)			
Crop				
Potatoes	17.75[1]	.178	1.00	$/cwt.
Cotton	17.22	.0 5	.08	$/lb.
Tobacco	15.19	.012	.08	$/lb.
Wheat	12.26	.130	1.06	$/bu.
Barley	11.34[a,2]	.068	.60	$/bu.
Corn	11.19	.061	.55	$/bu.
Tame Hay	10.49	1.370	13.07	$/ton
Buckwheat	8.60	.056	.65	$/bu.
Sweet Potatoes	8.46[d]	.085	1.00	$/cwt.
Rye	7.97	.056	.70	$/bu.
Oats	7.22	.030	.41	$/bu.
Livestock				
Hogs	8.00	.350	4.38	$/Head
Mules	6.01	5.237	87.09	$/Head
Sheep	4.23	.118	2.78	$/Head
All Cattle	4.19	.826	19.73	$/Head
Milk Cows	3.63	.843	23.23	$/Head
Horses	3.16	2.184	69.03	$/Head

[1]This estimate is based on 28 observations, 1882-1909.

[2]This estimate is based on 31 observations, 1879-1909.

Selected Crops and Livestock, Virginia: Ranking By
Yield and Number on Farms Random Variability Coefficients

	Random Variability Coefficient	Standard Deviation	Mean	Unit
	(Percent)			
Crop				
Barley	19.49[1]	3.888	19.95	Bu./acre
Wheat	17.11	1.751	10.23	Bu./acre
Buckwheat	15.96	1.709	10.71	Bu./acre
Corn	14.35	2.715	18.92	Bu./acre
Oats	12.51	1.452	11.60	Bu./acre
Tame Hay	12.19	.116	.95	Ton/acre
Tobacco	11.22	69.375	618.11	Lb./acre
Potatoes	10.69	4.676	43.73	Cwt./acre
Sweet Potatoes	10.64[e]	5.629	52.91	Cwt./acre
Rye	10.62	.861	8.11	Bu./acre
Livestock				
Hogs	2.04	16.288	799.5	1000's of Head
Mules	1.86	.723	38.9	1000's of Head
Horses	1.48	3.603	243.5	1000's of Head
Sheep	1.22	5.416	443.8	1000's of Head
All Cattle	.82	5.837	707.8	1000's of Head
Milk Cows	.56[a]	1.409	251.3	1000's of Head

[1]This estimate is based on 31 observations, 1879-1909.

TABLE A72

Selected Crops, Virginia: Ranking By
Gross Income Random Variability Coefficients

Crop	Random Variability Coefficient	Standard Deviation	Mean	Unit
	(Percent)			
Buckwheat	18.31[a]	1.262	6.89	$/acre
Tame Hay	15.09	1.851	12.27	$/acre
Potatoes	13.79	6.051	43.87	$/acre
Barley	13.78[1]	1.612	11.70	$/acre
Wheat	13.76	1.462	10.62	$/acre
Oats	12.67	.606	4.78	$/acre
Rye	12.67	.709	6.47	$/acre
Tobacco	12.64	6.086	48.14	$/acre
Sweet Potatoes	9.78[a,e]	5.518	56.42	$/acre
Corn	9.38	.961	10.25	$/acre

[1]This estimate is based on 31 observations, 1879-1909.

Selected Crops and Livestock, West Virginia: Ranking
By Price Random Variability Coefficients[1]

	Random Variability Coefficient	Standard Deviation	Mean	Unit
	(Percent)			
Crop				
Potatoes	25.90	.255	.98	$/cwt.
Barley	20.93[2]	.139	.66	$/bu.
Tobacco	17.96	.018	.10	$/lb.
Wheat	15.31	.156	1.02	$/bu.
Sweet Potatoes	13.25[d]	.179	1.35	$/cwt.
Corn	12.38	.069	.56	$/bu.
Oats	11.01	.044	.40	$/bu.
Tame Hay	10.34	1.188	11.49	$/ton
Rye	10.00	.074	.74	$/bu.
Buckwheat	7.64	.053	.69	$/bu.
Livestock				
Hogs	8.83	.408	4.64	$/Head
Horses	4.61	2.966	64.32	$/Head
All Cattle	4.20	.977	23.27	$/Head
Sheep	4.03	.107	2.66	$/Head
Mules	2.78	2.026	72.93	$/Head
Milk Cows	2.56	.693	27.10	$/Head

[1]All estimates are based on 43 observations unless otherwise noted, 1867–1909.

[2]This estimate is based on 15 observations, 1878–1892.

Selected Crops and Livestock, West Virginia: Ranking By
Yield and Number on Farms Random Variability Coefficients[1]

	Random Variability Coefficient	Standard Deviation	Mean	Unit
	(Percent)			
Crop				
Sweet Potatoes	22.86[e]	8.829	38.62	Cwt./acre
Rye	22.14	1.707	7.71	Bu./acre
Potatoes	18.66	8.071	43.26	Cwt./acre
Oats	14.37	2.897	20.16	Bu./acre
Wheat	13.43	1.442	10.74	Bu./acre
Buckwheat	11.97	1.326	11.08	Bu./acre
Tame Hay	11.89	.114	.96	Ton/acre
Corn	11.30	2.648	23.44	Bu./acre
Tobacco	8.56	51.010	596.26	Lb./acre
Livestock				
Mules	3.81	.292	7.7	1000's of Head
Hogs	2.30	7.714	336.1	1000's of Head
Sheep	1.87	11.081	593.8	1000's of Head
All Cattle	1.12	5.572	497.7	1000's of Head
Milk Cows	.53	.897	170.0	1000's of Head
Horses	.45	.656	146.6	1000's of Head

[1] All estimates are based on 43 observations, 1867–1909.

TABLE A75

Selected Crops, West Virginia: Ranking By
Gross Income Random Variability Coefficients[1]

Crop	Random Variability Coefficient	Standard Deviation	Mean	Unit
	(Percent)			
Rye	24.17	1.351	5.59	$/acre
Tobacco	16.06	9.458	58.91	$/acre
Sweet Potatoes	16.04[e]	8.184	51.02	$/acre
Potatoes	14.00	5.882	42.01	$/acre
Oats	12.74	1.013	7.95	$/acre
Wheat	12.35	1.341	10.86	$/acre
Buckwheat	11.24	.849	7.55	$/acre
Tame Hay	9.56	1.043	10.91	$/acre
Corn	5.83	.754	12.93	$/acre

[1] All estimates are based on 43 observations, 1867–1909.

Selected Crops and Livestock, North Carolina: Ranking
By Price Random Variability Coefficients

	Random Variability Coefficient	Standard Deviation	Mean	Unit
	(Percent)			
Crop				
Barley	18.63[c],[1]	.140	.75	$/bu.
Tobacco	15.40	.016	.10	$/lb.
Corn	11.39[d]	.071	.62	$/bu.
Sweet Potatoes	10.59	.086	.82	$/cwt.
Tame Hay	10.52	1.260	11.98	$/ton
Wheat	10.16[f]	.116	1.14	$/bu.
Cotton	9.63	.010	.10	$/lb.
Oats	9.47	.049	.52	$/bu.
Potatoes	9.33	.104	1.11	$/cwt.
Buckwheat	7.90	.050	.64	$/bu.
Rye	6.37	.055	.87	$/bu.
Livestock				
Milk Cows	9.57	1.736	18.14	$/Head
Sheep	6.69[a]	.111	1.66	$/Head
Hogs	6.22	.249	4.00	$/Head
Mules	5.67	4.991	87.97	$/Head
All Cattle	4.94	.651	13.18	$/Head
Horses	3.68	2.814	76.39	$/Head

[1]This estimate is based on 12 observations, 1881–1892.

Selected Crops and Livestock, North Carolina: Ranking By
Yield and Number on Farms Random Variability Coefficients

	Random Variability Coefficient	Standard Deviation	Mean	Unit
	(Percent)			
Crop				
Rye	20.67	1.044	5.05	Bu./acre
Wheat	20.67	1.334	6.46	Bu./acre
Oats	15.68	1.894	12.08	Bu./acre
Cotton	12.84	24.072	187.50	Lb./acre
Corn	12.53	1.590	12.69	Bu./acre
Buckwheat	11.33	.944	8.33	Bu./acre
Sweet Potatoes	11.18[e]	5.540	49.55	Cwt./acre
Tame Hay	11.11	.107	.96	Ton/acre
Tobacco	9.99	50.184	502.18	Lb./acre
Potatoes	8.78	3.811	43.39	Cwt./acre
Livestock				
Hogs	2.85	40.043	1407.3	1000's of Head
Sheep	1.56	5.386	346.3	1000's of Head
All Cattle	1.09	6.754	622.3	1000's of Head
Mules	.80	.813	101.3	1000's of Head
Horses	.76	1.035	136.2	1000's of Head
Milk Cows	.64	1.402	219.3	1000's of Head

TABLE A78

Selected Crops, North Carolina: Ranking By
Gross Income Random Variability Coefficients

Crop	Random Variability Coefficient	Standard Deviation	Mean	Unit
	(Percent)			
Wheat	18.57	1.347	7.25	$/acre
Tobacco	17.81	8.895	49.94	$/acre
Rye	17.50	.757	4.32	$/acre
Cotton	16.98[f]	3.179	18.72	$/acre
Oats	16.03	.998	6.23	$/acre
Buckwheat	13.46	.712	5.29	$/acre
Sweet Potatoes	12.61[d]	5.035	39.94	$/acre
Potatoes	11.88	5.736	48.29	$/acre
Tame Hay	11.76	1.352	11.49	$/acre
Corn	10.48	.822	7.48	$/acre

TABLE A79

Selected Crops and Livestock, South Carolina: Ranking
By Price Random Variability Coefficients

	Random Variability Coefficient	Standard Deviation	Mean	Unit
	(Percent)			
Crop				
Tame Hay	18.18	2.707	14.89	$/ton
Corn	13.44	.103	.77	$/bu.
Oats	12.90	.084	.65	$/bu.
Cotton	11.63[f]	.012	.10	$/lb.
Potatoes	11.24	.178	1.58	$/cwt.
Wheat	10.84	.149	1.37	$/bu.
Rye	8.50	.105	1.23	$/bu.
Sweet Potatoes	6.71[d]	.061	.91	$/cwt.
Livestock				
Hogs	10.32	.447	4.34	$/Head
All Cattle	6.63	.995	15.01	$/Head
Sheep	6.59	.122	1.85	$/Head
Milk Cows	6.11	1.276	20.88	$/Head
Mules	5.37	5.359	99.87	$/Head
Horses	3.17	2.762	87.22	$/Head

TABLE A80

Selected Crops and Livestock, South Carolina: Ranking By
Yield and Number on Farms Random Variability Coefficients

	Random Variability Coefficient	Standard Deviation	Mean	Unit
	(Percent)			
Crop				
Sweet Potatoes	23.29[e]	8.866	38.07	Cwt./acre
Wheat	18.89	1.159	6.13	Bu./acre
Corn	18.36	1.883	10.26	Bu./acre
Cotton	15.01	26.512	176.66	Lb./acre
Oats	14.74	1.911	12.96	Bu./acre
Potatoes	14.21	6.813	47.96	Cwt./acre
Rye	14.03	.635	4.52	Bu./acre
Tobacco	12.91[1]	94.306	730.24	Lb./acre
Tame Hay	8.59	.084	.98	Ton/acre

	Random Variability Coefficient	Standard Deviation	Mean	Unit
	(Percent)			
Livestock				
Sheep	2.77	2.437	87.1	1000's of Head
All Cattle	.66	2.111	319.7	1000's of Head
Mules	.53	.471	88.1	1000's of Head
Milk Cows	.46	.563	121.2	1000's of Head
Horses	.45	.288	63.6	1000's of Head
Hogs	.42	19.167	543.7	1000's of Head

[1]This estimate is based on 21 observations, 1889-1909.

TABLE A81

Selected Crops, South Carolina: Ranking By
Gross Income Random Variability Coefficients

Crop	Random Variability Coefficient	Standard Deviation	Mean	Unit
	(Percent)			
Tame Hay	23.09	3.374	14.62	$/acre
Tobacco	22.21[1]	13.768	61.99	$/acre
Sweet Potatoes	20.74[d]	7.409	35.73	$/acre
Cotton	20.66[f]	3.673	17.77	$/acre
Potatoes	18.24	13.859	75.98	$/acre
Wheat	17.83	1.474	8.27	$/acre
Rye	16.09	.881	5.48	$/acre
Corn	12.96	1.004	7.74	$/acre
Oats	8.52	.710	8.33	$/acre

[1]This estimate is based on 21 observations, 1889-1909.

Selected Crops and Livestock, Georgia: Ranking
By Price Random Variability Coefficients

	Random Variability Coefficient	Standard Deviation	Mean	Unit
	(Percent)			
Crop				
Corn	12.04	.087	.72	$/bu.
Potatoes	11.01	.178	1.62	$/cwt.
Oats	10.41	.066	.64	$/bu.
Sweet Potatoes	9.76[d]	.090	.92	$/cwt.
Tame Hay	9.55	1.496	15.67	$/ton
Wheat	8.89	.112	1.26	$/bu.
Rye	7.86[1]	.094	1.20	$/bu.
Cotton	7.77[f]	.008	.10	$/lb.
Barley	7.63[c,2]	.063	.82	$/bu.
Livestock				
Hogs	7.72	.305	3.96	$/Head
All Cattle	6.40	.841	13.14	$/Head
Milk Cows	5.45	1.075	19.73	$/Head
Horses	5.19	4.305	82.96	$/Head
Mules	4.81	4.815	100.04	$/Head
Sheep	4.32	.072	1.66	$/Head

[1] This estimate is based on 43 observations, 1867-1909.

[2] This estimate is based on 13 observations, 1830-1893.

Selected Crops and Livestock, Georgia: Ranking By
Yield and Number on Farms Random Variability Coefficients

	Random Variability Coefficient	Standard Deviation	Mean	Unit
	(Percent)			
Crop				
Wheat	20.38	1.289	6.32	Bu./acre
Rye	17.65[1]	.716	4.06	Bu./acre
Sweet Potatoes	14.12[e]	6.279	44.48	Cwt./acre
Corn	13.01	1.364	10.48	Bu./acre
Cotton	12.98	20.168	155.34	Lb./acre
Oats	10.95	1.416	12.93	Bu./acre
Tobacco	10.27[c,2]	66.643	648.64	Lb./acre
Potatoes	8.66	3.848	44.43	Cwt./acre
Tame Hay	8.12	.069	.85	Ton/acre
Livestock				
Hogs	1.48	20.685	1400.6	1000's of Head
Sheep	1.21	4.704	390.3	1000's of Head
All Cattle	.75	6.631	883.6	1000's of Head
Milk Cows	.58	1.615	280.0	1000's of Head
Mules	.56	.903	162.2	1000's of Head
Horses	.36	.380	106.9	1000's of Head

[1]This estimate is based on 43 observations, 1867-1909.
[2]This estimate is based on 11 observations, 1899-1909.

TABLE A84

Selected Crops, Georgia: Ranking By
Gross Income Random Variability Coefficients

Crop	Random Variability Coefficient	Standard Deviation	Mean	Unit
	(Percent)			
Wheat	34.05	2.673	7.85	$/acre
Rye	25.35	1.230	4.85	$/acre
Sweet Potatoes	17.02[a,d]	7.132	41.91	$/acre
Tobacco	15.29[c,1]	24.238	158.53	$/acre
Potatoes	14.92	10.693	71.69	$/acre
Oats	14.37	1.174	8.17	$/acre
Corn	12.17	.903	7.42	$/acre
Cotton	11.50[b,f]	1.786	15.52	$/acre
Tame Hay	8.11	1.398	13.16	$/acre

This estimate is based on 11 observations, 1899-1909.

Selected Crops and Livestock, Florida: Ranking
By Price Random Variability Coefficients

	Random Variability Coefficient	Standard Deviation	Mean	Unit
	(Percent)			
Crop				
Cotton	26.23[f]	.030	.11	$/lb.
Potatoes	19.63[1]	.341	1.74	$/cwt.
Oats	10.48[a]	.080	.76	$/bu.
Corn	9.66	.080	.83	$/bu.
Livestock				
Hogs	11.24	.304	2.70	$/Head
Sheep	6.56	.125	1.90	$/Head
Mules	6.39	6.401	100.22	$/Head
All Cattle	5.70	.541	9.49	$/Head
Milk Cows	5.33	.914	17.13	$/Head
Horses	4.30	3.360	78.10	$/Head

[1] This estimate is based on 29 observations, 1881-1909.

TABLE A86

Selected Crops and Livestock, Florida: Ranking by
Yield and Number on Farms Random Variability Coefficients

	Random Variability Coefficient	Standard Deviation	Mean	Unit
	(Percent)			
Crop				
Cotton	18.94	21.748	114.84	Lb./acre
Oats	14.03	1.575	11.23	Bu./acre
Tame Hay	12.83[1]	.086	.67	Ton/acre
Potatoes	10.47[2]	4.183	39.97	Cwt./acre
Corn	7.87[3]	.756	9.60	Bu./acre
Tobacco	5.92	41.996	709.39	Lb./acre
Sweet Potatoes	5.22[e]	2.689	51.48	Cwt./acre
Livestock				
Hogs	2.45	10.906	444.7	1000's of Head
Mules	2.13[b]	.263	12.3	1000's of Head
Sheep	2.13	2.204	103.6	1000's of Head
Milk Cows	1.46	.960	65.6	1000's of Head
All Cattle	1.02	6.318	620.5	1000's of Head
Horses	.66[b]	.203	30.5	1000's of Head

[1] This estimate is based on 21 observations, 1889-1909.
[2] This estimate is based on 29 observations, 1881-1909.
[3] This estimate is based on 13 observations, 1897-1909.

Selected Crops, Florida: Ranking By
Gross Income Random Variability Coefficients

Crop	Random Variability Coefficient	Standard Deviation	Mean	Unit
	(Percent)			
Cotton	29.90[f]	3.787	12.97	$/acre
Tame Hay	22.53[1]½	2.342	10.40	$/acre
Potatoes	18.13[2]	12.806	70.62	$/acre
Oats	16.76	1.416	8.45	$/acre
Corn	10.17	.819	8.05	$/acre

[1]This estimate is based on 21 observations, 1889–1909.
[2]This estimate is based on 29 observations, 1881–1909.

TABLE A88

Selected Crops and Livestock, Kentucky: Ranking
By Price Random Variability Coefficients

	Random Variability Coefficient	Standard Deviation	Mean	Unit
	(Percent)			
Crop				
Potatoes	29.45	.290	.98	$/cwt.
Corn	24.61	.109	.44	$/bu.
Tobacco	21.61	.016	.07	$/lb.
Wheat	17.66	.170	.96	$/bu.
Sweet Potatoes	14.92[d]	.171	1.15	$/cwt.
Oats	13.12	.050	.38	$/bu.
Buckwheat	11.15[c,1]	.073	.66	$/bu.
Barley	10.64	.077	.72	$/bu.
Tame Hay	10.52	1.214	11.54	$/ton
Rye	8.93	.066	.73	$/bu.
Livestock				
Hogs	10.53	.467	4.44	$/Head
Sheep	7.06	.197	2.79	$/Head
All Cattle	5.64	1.303	23.10	$/Head
Milk Cows	3.77	1.025	27.21	$/Head
Horses	2.85	1.824	63.98	$/Head
Mules	2.36	1.660	70.27	$/Head

[1]This estimate is based on 12 observations, 1881–1892.

Selected Crops and Livestock, Kentucky: Ranking By
Yield and Number on Farms Random Variability Coefficients

	Random Variability Coefficient	Standard Deviation	Mean	Unit
	(Percent)			
Crop				
Rye	23.99	2.077	8.66	Bu./acre
Wheat	21.80	2.324	10.66	Bu./acre
Potatoes	20.74	9.330	44.98	Cwt./acre
Sweet Potatoes	18.41[e]	7.136	38.76	Cwt./acre
Corn	13.69	3.397	24.81	Bu./acre
Oats	13.51	2.317	17.15	Bu./acre
Barley	11.39	2.827	24.82	Bu./acre
Tobacco	10.33	84.536	818.21	Lb./acre
Tame Hay	8.56	.087	1.01	Ton/acre
Livestock				
Hogs	3.00	55.581	1851.9	1000's of Head
All Cattle	1.83[b]	17.291	945.7	1000's of Head
Sheep	1.09	9.716	892.9	1000's of Head
Milk Cows	.67	2.088	309.6	1000's of Head
Mules	.38	.563	146.8	1000's of Head
Horses	.17	.670	393.5	1000's of Head

TABLE A90

Selected Crops, Kentucky: Ranking By
Gross Income Random Variability Coefficients

Crop	Random Variability Coefficient	Standard Deviation	Mean	Unit
	(Percent)			
Rye	33.84	2.133	6.30	$/acre
Wheat	29.30	2.881	9.83	$/acre
Barley	18.09	3.236	17.89	$/acre
Tobacco	17.45	10.546	60.43	$/acre
Sweet Potatoes	14.93[d]	7.293	48.83	$/acre
Potatoes	12.59	5.363	42.59	$/acre
Oats	9.33	.609	6.52	$/acre
Tame Hay	9.31	1.082	11.63	$/acre
Corn	8.50	.915	10.76	$/acre

Selected Crops and Livestock, Tennessee: Ranking
By Price Random Variability Coefficients

	Random Variability Coefficient	Standard Deviation	Mean	Unit
	(Percent)			
Crop				
Rye	26.43	.215	.81	$/bu.
Corn	25.58	.121	.47	$/bu.
Potatoes	23.67	.237	1.00	$/cwt.
Tobacco	23.12	.019	.08	$/lb.
Sweet Potatoes	18.74[d]	.184	.98	$/cwt.
Barley	14.68	.108	.73	$/bu.
Wheat	13.25	.131	.99	$/bu.
Oats	12.20	.050	.41	$/bu.
Buckwheat	9.51	.070	.74	$/bu.
Cotton	9.49[f]	.009	.10	$/lb.
Tame Hay	9.44[b]	1.212	12.84	$/ton
Livestock				
Hogs	12.28	.502	4.09	$/Head
All Cattle	6.32	.969	15.34	$/Head
Milk Cows	5.46	1.142	20.90	$/Head
Sheep	3.54	.074	2.09	$/Head
Horses	3.51	2.360	67.30	$/Head
Mules	2.90	2.182	75.18	$/Head

Selected Crops and Livestock, Tennessee: Ranking By
Yield and Number on Farms Random Variability Coefficients

	Random Variability Coefficient	Standard Deviation	Mean	Unit
	(Percent)			
Crop				
Wheat	25.96	2.124	8.18	Bu./acre
Sweet Potatoes	18.29	7.529[e]	41.17	Cwt./acre
Tobacco	17.48	115.890	662.98	Lb./acre
Cotton	16.80	31.856	189.59	Lb./acre
Oats	15.33	2.354	15.35	Bu./acre
Barley	14.44	2.519	17.45	Bu./acre
Rye	13.56	.749	5.52	Bu./acre
Buckwheat	13.21	.972	7.36	Bu./acre
Corn	11.60	2.527	21.79	Bu./acre
Potatoes	10.88	4.248	39.05	Cwt./acre
Tame Hay	6.04	.063	1.05	Ton/acre
Livestock				
Hogs	1.73	32.120	1853.8	1000's of Head
Milk Cows	1.41	4.004	283.1	1000's of Head
Sheep	.81	4.564	560.3	1000's of Head
Mules	.53	1.018	193.2	1000's of Head
All Cattle	.37	3.279	880.8	1000's of Head
Horses	.30	.904	300.6	1000's of Head

TABLE A93

Selected Crops, Tennessee: Ranking By
Gross Income Random Variability Coefficients

Crop	Random Variability Coefficient	Standard Deviation	Mean	Unit
	(Percent)			
Wheat	35.24	2.778	7.89	$/acre
Rye	26.51	1.170	4.41	$/acre
Tobacco	23.06	12.849	55.72	$/acre
Barley	20.01	2.609	13.04	$/acre
Sweet Potatoes	18.23[d]	7.703	42.25	$/acre
Buckwheat	15.21	.820	5.39	$/acre
Cotton	14.66[f]	2.746	18.74	$/acre
Oats	13.97	.880	6.30	$/acre
Potatoes	12.54	4.853	38.71	$/acre
Tame Hay	11.26	1.517	13.47	$/acre
Corn	8.53[b]	.870	10.20	$/acre

Selected Crops and Livestock, Alabama: Ranking
By Price Random Variability Coefficients

	Random Variability Coefficient	Standard Deviation	Mean	Unit
	(Percent)			
Crop				
Corn	12.57	.085	.68	$/bu.
Wheat	10.54	.124	1.18	$/bu.
Tame Hay	10.35	1.470	14.21	$/ton
Rye	10.15[c,1]	.110	1.09	$/bu.
Cotton	9.99[f]	.010	.10	$/lb.
Oats	8.77	.055	.63	$/bu.
Potatoes	8.58	.138	1.60	$/cwt
Barley	8.44[c,2]	.070	.83	$/bu.
Sweet Potatoes	8.38[d]	.077	.92	$/cwt.
Livestock				
Hogs	9.81	.356	3.63	$/Head
Sheep	7.46	.121	1.62	$/Head
Horses	6.29	4.367	69.41	$/Head
All Cattle	6.15	.736	11.97	$/Head
Milk Cows	5.28	.923	17.46	$/Head
Mules	3.92	3.302	84.29	$/Head

[1] This estimate is based on 30 observations, 1880–1909.
[2] This estimate is based on 12 observations, 1881–1892.

Selected Crops and Livestock, Alabama: Ranking By
Yield and Number on Farms Random Variability Coefficients

	Random Variability Coefficient	Standard Deviation	Mean	Unit
	(Percent)			
Crop				
Wheat	20.56	1.353	6.58	Bu./acre
Cotton	15.53	22.455	144.55	Lb./acre
Sweet Potatoes	14.95[e]	6.637	44.38	Cwt./acre
Oats	12.53	1.554	12.40	Bu./acre
Corn	12.12	1.552	12.80	Bu./acre
Tame Hay	8.72	.080	.92	Ton/acre
Potatoes	7.65	3.583	46.84	Cwt./acre
Livestock				
Hogs	1.82	22.298	1222.8	1000's of Head
Milk Cows	.98	2.448	248.6	1000's of Head
Sheep	.78	2.166	278.2	1000's of Head
All Cattle	.69	5.532	798.1	1000's of Head
Mules	.61	.879	144.4	1000's of Head
Horses	.43	.512	119.2	1000's of Head

TABLE A96

Selected Crops, Alabama: Ranking By
Gross Income Random Variability Coefficients

Crop	Random Variability Coefficient	Standard Deviation	Mean	Unit
	(Percent)			
Wheat	19.11	1.467	7.68	$/acre
Tame Hay	13.75	1.787	13.00	$/acre
Sweet Potatoes	13.19[d]	5.408	41.00	$/acre
Corn	12.30	1.052	8.56	$/acre
Oats	12.29	.948	7.72	$/acre
Cotton	11.73[f]	1.698	14.48	$/acre
Potatoes	9.06	6.820	75.31	$/acre

Selected Crops and Livestock, Mississippi: Ranking
By Price Random Variability Coefficients

	Random Variability Coefficient	Standard Deviation	Mean	Unit
	(Percent)			
Crop				
Rye	15.62[c],[1]	.161	1.03	$/bu.
Corn	14.20	.097	.68	$/bu.
Potatoes	13.82	.210	1.52	$/cwt.
Oats	10.36	.068	.65	$/bu.
Cotton	9.51[f]	.010	.10	$/lb.
Sweet Potatoes	9.14[d]	.090	.98	$/cwt.
Tame Hay	7.30	1.002	13.74	$/ton
Wheat	7.16	.087	1.21	$/bu.
Livestock				
Hogs	7.72	.278	3.61	$/Head
Sheep	6.91[a]	.113	1.64	$/Head
Milk Cows	5.66	1.047	18.49	$/Head
Horses	4.20	2.859	68.10	$/Head
Mules	3.86	3.348	86.85	$/Head
All Cattle	3.83	.487	12.71	$/Head

[1]This estimate is based on 12 observations, 1881–1892.

Selected Crops and Livestock, Mississippi: Ranking By
Yield and Number on Farms Random Variability Coefficients

	Random Variability Coefficient	Standard Deviation	Mean	Unit
	(Percent)			
Crop				
Wheat	18.86	1.409	7.47	Bu./acre
Cotton	17.09	31.455	184.00	Lb./acre
Corn	14.58	2.111	14.48	Bu./acre
Sweet Potatoes	11.40[e]	5.286	46.37	Cwt./acre
Oats	11.31	1.507	13.32	Bu./acre
Potatoes	9.18	3.858	42.05	Cwt./acre
Tame Hay	4.72	.058	1.22	Ton/acre
Livestock				
Sheep	2.42	6.722	278.0	1000's of Head
Hogs	2.06	21.955	1066.7	1000's of Head
Milk Cows	.63	1.643	260.1	1000's of Head
All Cattle	.62	4.992	804.9	1000's of Head
Horses	.54	.838	155.1	1000's of Head
Mules	.43	.691	159.9	1000's of Head

TABLE A99

Selected Crops, Mississippi: Ranking By
Gross Income Random Variability Coefficients

Crop	Random Variability Coefficient	Standard Deviation	Mean	Unit
	(Percent)			
Wheat	17.80	1.578	8.86	$/acre
Sweet Potatoes	17.05[d]	7.752	45.47	$/acre
Corn	13.28	1.279	9.63	$/acre
Cotton	12.76[f]	2.383	18.67	$/acre
Potatoes	12.55	8.029	63.98	$/acre
Oats	9.06	.772	8.51	$/acre
Tame Hay	6.33	1.049	16.57	$/acre

Selected Crops and Livestock, Arkansas: Ranking
By Price Random Variability Coefficients

	Random Variability Coefficient	Standard Deviation	Mean	Unit
	(Percent)			
Crop				
Potatoes	26.02	.340	1.31	$/cwt.
Corn	22.20	.127	.57	$/bu.
Tobacco	18.94	.023	.12	$/lb.
Rye	14.88	.140	.94	$/bu.
Wheat	13.10	.135	1.03	$/bu.
Sweet Potatoes	12.46[d]	.128	1.03	$/cwt.
Oats	12.19	.062	.51	$/bu.
Tame Hay	10.42	1.262	12.11	$/ton
Cotton	10.04[f]	.010	.10	$/lb.
Livestock				
Hogs	16.54	.532	3.22	$/Head
Sheep	8.32[a]	.148	1.78	$/Head
Horses	6.39	3.553	55.61	$/Head
Mules	5.75	4.061	70.65	$/Head
Milk Cows	4.79	.828	17.29	$/Head
All Cattle	4.42	.563	12.73	$/Head

Selected Crops and Livestock, Arkansas: Ranking By
Yield and Number on Farms Random Variability Coefficients

	Random Variability Coefficient	Standard Deviation	Mean	Unit
	(Percent)			
Crop				
Sweet Potatoes	27.67[e]	12.618	45.60	Cwt./acre
Wheat	20.87	1.635	7.83	Bu./acre
Cotton	17.54	37.805	215.59	Lb./acre
Corn	14.46	2.709	18.73	Bu./acre
Rye	14.34	.883	6.16	Bu./acre
Oats	13.73	2.513	18.31	Bu./acre
Potatoes	13.14	6.093	46.39	Cwt./acre
Tame Hay	10.84	.117	1.08	Ton/acre
Tobacco	9.50	46.031	484.41	Lb./acre
Livestock				
Hogs	5.71	74.164	1297.7	1000's of Head
All Cattle	1.82	13.257	728.8	1000's of Head
Mules	1.44	1.784	124.1	1000's of Head
Sheep	1.33	2.560	192.9	1000's of Head
Milk Cows	1.13	2.904	257.5	1000's of Head
Horses	1.12	2.037	182.5	1000's of Head

TABLE A102

Selected Crops, Arkansas: Ranking By
Gross Income Random Variability Coefficients

Crop	Random Variability Coefficient	Standard Deviation	Mean	Unit
	(Percent)			
Wheat	25.67	2.088	8.13	$/acre
Tobacco	20.43	11.776	57.64	$/acre
Sweet Potatoes	19.98[a,d]	9.646	48.27	$/acre
Rye	18.58	1.080	5.81	$/acre
Cotton	17.94[f]	3.760	20.95	$/acre
Potatoes	16.63	10.024	60.28	$/acre
Oats	13.80	1.258	9.12	$/acre
Corn	13.01	1.365	10.49	$/acre
Tame Hay	10.94	1.426	13.03	$/acre

Selected Crops and Livestock, Louisiana: Ranking
By Price Random Variability Coefficients

	Random Variability Coefficient	Standard Deviation	Mean	Unit
	(Percent)			
Crop				
Tame Hay	21.62[c,1]	4.307	19.93	$/ton
Potatoes	20.93	.318	1.52	$/cwt.
Sweet Potatoes	19.40[d]	.187	.97	$/cwt.
Oats	18.56[2]	.123	.66	$/bu.
Rye	17.82[c,3]	.160	.90	$/bu.
Corn	14.67	.101	.69	$/bu.
Cotton	10.13[f]	.010	.10	$/1b.
Livestock				
Milk Cows	7.61[a]	1.540	20.23	$/Head
Horses	6.74	4.016	59.62	$/Head
Hogs	6.65	.250	3.76	$/Head
All Cattle	4.84	.636	13.13	$/Head
Mules	4.71	4.143	87.98	$/Head
Sheep	4.42	.077	1.73	$/Head

[1]This estimate is based on 10 observations, 1866-1875.
[2]This estimate is based on 43 observations, 1867-1909.
[3]This estimate is based on 12 observations, 1881-1892.

Selected Crops and Livestock, Louisiana: Ranking By
Yield and Number on Farms Random Variability Coefficients

	Random Variability Coefficient	Standard Deviation.	Mean	Unit
	(Percent)			
Crop				
Cotton	24.00	53.114	221.27	Lb./acre
Potatoes	20.79	6.786	32.64	Cwt./acre
Oats	17.94	2.591	14.44	Bu./acre
Sweet Potatoes	16.00[e]	6.700	41.88	Cwt./acre
Corn	11.63	1.752	15.06	Bu./acre
Tame Hay	6.67[1]	.093	1.40	Ton/acre
Livestock				
Hogs	3.03	19.056	628.8	1000's of Head
Sheep	1.81	2.789	154.5	1000's of Head
Milk Cows	1.52[a]	2.290	150.6	1000's of Head
All Cattle	.67	4.086	610.5	1000's of Head
Mules	.62	.615	99.6	1000's of Head
Horses	.56	.754	134.7	1000's of Head

[1] This estimate is based on 29 observations, 1881-1909.

TABLE A105

Selected Crops, Louisiana: Ranking
Gross Income Random Variability Coefficients

Crop	Random Variability Coefficient	Standard Deviation	Mean	Unit
	(Percent)			
Potatoes	31.72	15.741	49.62	$/acre
Sweet Potatoes	22.94[d]	9.095	39.65	$/acre
Oats	17.68[a,1]	1.688	9.55	$/acre
Cotton	17.13[f]	3.833	22.38	$/acre
Corn	12.16	1.240	10.19	$/acre
Tame Hay	10.49[c,2]	1.611	15.35	$/acre

[1] This estimate is based on 43 observations, 1867-1909.
[2] This estimate is based on 29 observations, 1881-1909.

Selected Crops and Livestock, Oklahoma: Ranking
By Price Random Variability Coefficients[c]

	Random Variability Coefficient	Standard Deviation	Mean	Unit
	(Percent)			
Crop				
Corn	34.27[1]	.142	.42	$/bu.
Potatoes	21.38[2]	.294	1.38	$/cwt.
Sweet Potatoes	18.41[2]	.246	1.34	$/cwt.
Cotton	17.78[3]	.014	.08	$/lb.
Wheat	16.71[3]	.113	.68	$/bu.
Rye	12.72[1]	.082	.65	$/bu.
Barley	12.41[1]	.056	.45	$/bu.
Livestock[4]				
Hogs	14.47	.785	5.43	$/Head
Sheep	9.59	.248	2.58	$/Head
All Cattle	7.89	1.496	18.96	$/Head
Milk Cows	6.44	1.583	24.56	$/Head
Mules	5.23	3.056	58.44	$/Head
Horses	3.10	1.304	42.02	$/Head

[1] These estimates are based on 11 observations, 1899-1909.
[2] These estimates are based on 14 observations, 1896-1909.
[3] These estimates are based on 16 observations, 1894-1909.
[4] All livestock estimates are based on 17 observations, 1893-1909.

Selected Crops and Livestock, Oklahoma: Ranking By
Yield and Number on Farms Random Variability Coefficients

	Random Variability Coefficient	Standard Deviation	Mean	Unit
	(Percent)			
Crop				
Corn	29.93[1]	6.671	24.77	Bu./acre
Oats	23.58[2]	6.180	26.22	Bu./acre
Wheat	22.08[3]	2.772	12.56	Bu./acre
Cotton	21.71[3]	44.618	205.56	Lb./acre
Rye	21.33[1]	2.062	9.66	Bu./acre
Barley	19.71[c,1]	3.262	16.55	Bu./acre
Sweet Potatoes	18.86[4]	10.011	53.07	Cwt./acre
Potatoes	18.07[4]	8.196	45.36	Cwt./acre
Tame Hay	13.49[1]	.226	1.67	Ton/acre
Livestock[5]				
Hogs	6.27	60.074	958.2	1000's of Head
Sheep	3.14	1.874	59.6	1000's of Head
Mules	1.65	2.150	130.6	1000's of Head
All Cattle	1.08	22.099	2041.8	1000's of Head
Horses	1.04	5.876	567.4	1000's of Head
Milk Cows	.97	3.089	319.9	1000's of Head

[1] These estimates are based on 11 observations, 1899-1909.
[2] This estimates are based on 13 observations, 1897-1909.
[3] These estimates are based on 16 observations, 1894-1909.
[4] These estimates are based on 14 observations, 1896-1909.
[5] All livestock estimates are based on 17 observations, 1893-1909.

TABLE A108

Selected Crops, Oklahoma: Ranking By
Gross Income Random Variability Coefficients[c]

Crop	Random Variability Coefficient	Standard Deviation	Mean	Unit
	(Percent)			
Wheat	27.42[1]	2.300	8.39	$/acre
Rye	22.62[2]	1.392	6.15	$/acre
Cotton	21.08[1]	3.384	16.05	$/acre
Sweet Potatoes	19.10[3]	13.399	70.16	$/acre
Potatoes	15.90[3]	9.787	61.54	$/acre
Corn	14.91[2]	1.418	9.51	$/acre
Barley	14.90[2]	1.075	7.21	$/acre

[1]These estimates are based on 16 observations, 1894–1909.
[2]These estimates are based on 11 observations, 1899–1909.
[3]These estimates are based on 14 observations, 1896–1909.

TABLE A109

Selected Crops and Livestock, Texas: Ranking
By Price Random Variability Coefficients

	Random Variability Coefficient	Standard Deviation	Mean	Unit
	(Percent)			
Crop				
Corn	32.93	.198	.60	$/bu.
Barley	25.21[a]	.199	.79	$/bu.
Sweet Potatoes	20.32[d]	.241	1.19	$/cwt.
Tame Hay	19.27	1.977	10.26	$/ton
Oats	16.79	.089	.53	$/bu.
Potatoes	16.00	.276	1.73	$/cwt.
Rye	15.26	.141	.93	$/bu.
Wheat	15.05	.159	1.06	$/bu.
Cotton	10.45[f]	.010	.10	$/lb.
Livestock				
Hogs	8.65	.312	3.61	$/Head
All Cattle	7.03	.812	11.56	$/Head
Sheep	6.27	.117	1.87	$/Head
Mules	5.84	3.081	52.75	$/Head
Milk Cows	4.87	.863	17.72	$/Head
Horses	3.96	1.348	34.08	$/Head

TABLE A110

Selected Crops and Livestock, Texas: Ranking By
Yield and Number on Farms Random Variability Coefficients

	Random Variability Coefficient	Standard Deviation	Mean	Unit
	(Percent)			
Crop				
Sweet Potatoes	24.67[e]	11.221	45.48	Cwt./acre
Wheat	24.57	2.702	11.00	Bu./acre
Barley	21.47	4.060	18.92	Bu./acre
Corn	21.07	4.255	20.19	Bu./acre
Cotton	20.78	40.857	196.66	Lb./acre
Oats	17.19	4.480	26.06	Bu./acre
Rye	16.11	1.718	10.67	Bu./acre
Tame Hay	12.89	.173	1.34	Ton/acre
Potatoes	12.68	5.196	40.98	Cwt./acre
Livestock				
Hogs	4.06	88.642	2180.7	1000's of Head
Sheep	3.23	105.948	3278.3	1000's of Head
Milk Cows	2.49	16.463	662.2	1000's of Head
All Cattle	2.09	149.326	7141.7	1000's of Head
Mules	1.20	3.667	304.6	1000's of Head
Horses	.56	6.146	1095.2	1000's of Head

TABLE A111

Selected Crops, Texas: Ranking By
Gross Income Random Variability Coefficients

Crop	Random Variability Coefficient	Standard Deviation	Mean	Unit
	(Percent)			
Barley	29.41[b]	4.473	15.21	$/acre
Wheat	24.43	2.762	11.30	$/acre
Sweet Potatoes	22.01[a,d]	11.982	54.44	$/acre
Potatoes	19.94	14.077	70.59	$/acre
Tame Hay	17.69	2.402	13.58	$/acre
Cotton	17.66[f]	3.299	18.69	$/acre
Rye	17.17	1.668	9.71	$/acre
Oats	15.03	2.054	13.62	$/acre
Corn	10.14	1.190	11.73	$/acre

Selected Crops and Livestock, Montana: Ranking
By Price Random Variability Coefficients[1]

	Random Variability Coefficient	Standard Deviation	Mean	Unit
	(Percent)			
Crop				
Potatoes	25.44	.250	.98	$/cwt.
Barley	19.16	.115	.60	$/bu.
Tame Hay	18.85	1.734	9.20	$/ton
Oats	12.53	.054	.43	$/bu.
Wheat	9.41	.071	.75	$/bu.
Rye	7.11[c,2]	.046	.65	$/bu.
Livestock				
Hogs	17.51	1.444	8.25	$/Head
Sheep	8.30	.217	2.62	$/Head
All cattle	7.76	1.690	21.78	$/Head
Milk Cows	7.72	2.641	34.20	$/Head
Horses	7.06	2.945	41.72	$/Head
Mules	4.93	2.822	57.20	$/Head

[1]All estimates are based on 28 observations unless otherwise noted, 1882-1909.
[2]This estimate is based on 11 observations, 1899-1909.

Selected Crops and Livestock, Montana: Ranking By
Yield and Number on Farms Random Variability Coefficients[1]

	Random Variability Coefficient	Standard Deviation	Mean	Unit
	(Percent)			
Crop				
Rye	19.14[c,2]	3.191	16.67	Bu./acre
Corn	16.19[3]	4.235	26.16	Bu./acre
Barley	14.72	5.072	34.45	Bu./acre
Wheat	14.33	3.452	24.10	Bu./acre
Potatoes	12.04	9.727	80.79	Cwt./acre
Tame Hay	9.27	.164	1.77	Ton/acre
Oats	8.16	2.918	35.76	Bu./acre
Livestock				
Mules	10.36	.196	1.9	1000's of Head
Sheep	3.76	131.811	3504.8	1000's of Head
Milk Cows	2.87	.899	31.3	1000's of Head
All Cattle	2.65	27.089	1021.3	1000's of Head
Hogs	2.63	.983	37.4	1000's of Head
Horses	1.15	2.718	236.8	1000's of Head

[1]All estimates are based on 28 observations unless noted, 1882–1909.
[2]This estimate is based on 11 observations, 1899–1909.
[3]This estimate is based on 18 observations, 1892–1909.

TABLE A114

Selected Crops, Montana: Ranking By
Gross Income Random Variability Coefficients

Crop	Random Variability Coefficient	Standard Deviation	Mean	Unit
	(Percent)			
Wheat	27.74	5.023	18.11	$/acre
Potatoes	24.91	19.575	78.59	$/acre
Corn	19.83[c,2]	3.667	18.50	$/acre
Tame Hay	17.63	2.848	16.16	$/acre
Barley	17.38	3.609	20.77	$/acre
Rye	16.28[c,3]	1.784	10.96	$/acre
Oats	12.14	1.869	15.39	$/acre

[1]All estimates are based on 28 observations unless otherwise noted, 1882–1909.
[2]This estimate is based on 18 observations, 1892–1909.
[3]This estimate is based on 11 observations, 1899–1909.

Selected Crops and Livestock, Idaho: Ranking
By Price Random Variability Coefficients[1]

	Random Variability Coefficient	Standard Deviation	Mean	Unit
	(Percent)			
Crop				
Tame Hay	29.56	2.182	7.38	$/ton
Potatoes	28.64	.261	.91	$/cwt.
Barley	25.73	.140	.54	$/bu.
Oats	11.27	.050	.44	$/bu.
Wheat	7.66	.054	.71	$/bu.
Livestock				
Horses	17.23	7.887	45.78	$/Head
Hogs	13.80	.943	6.84	$/Head
Sheep	12.12	.312	2.58	$/Head
Mules	8.49	4.965	58.48	$/Head
Milk Cows	7.82	2.415	30.88	$/Head
All Cattle	5.85	1.182	20.23	$/Head

[1]All estimates are based on 28 observations, 1882–1909.

TABLE A116

Selected Crops and Livestock, Idaho: Ranking By
Yield and Number on Farms Random Variability Coefficients[1]

	Random Variability Coefficient	Standard Deviation	Mean	Unit
	(Percent)			
Crop				
Potatoes	15.73	11.229	71.39	Cwt./acre
Barley	15.53	4.469	28.78	Bu./acre
Corn	13.70	3.844	28.05	Bu./acre
Rye	13.30	1.740	13.08	Bu./acre
Wheat	11.21	2.514	22.43	Bu./acre
Tame Hay	10.65	.207	1.95	Ton/acre
Oats	9.74	3.314	34.01	Bu./acre
Livestock				
Mules	10.77	.196	1.8	1000's of Head
Sheep	3.02	40.540	1343.6	1000's of Head
All Cattle	2.00	7.852	393.1	1000's of Head
Hogs	1.23	1.090	88.3	1000's of Head
Horses	.98	1.259	127.9	1000's of Head
Milk Cows	.71	.305	42.7	1000's of Head

[1]All estimates are based on 28 observations, 1882–1909.

Selected Crops, Idaho: Ranking By
Gross Income Random Variability Coefficients

Crop	Random Variability Coefficient	Standard Deviation	Mean	Unit
	(Percent)			
Tame Hay	31.27	4.493	14.37	$/acre
Potatoes	29.92	19.341	64.65	$/acre
Barley	28.95	4.577	15.81	$/acre
Rye	12.07[c,2]	1.019	8.44	$/acre
Oats	11.05	1.656	14.99	$/acre
Wheat	9.56	1.522	15.93	$/acre

[1]All estimates are based on 28 observations unless otherwise noted, 1882–1909.
[2]This estimate is based on 11 observations, 1899–1909.

TABLE A118

Selected Crops and Livestock, Wyoming: Ranking
By Price Random Variability Coefficients[1]

	Random Variability Coefficient	Standard Deviation	Mean	Unit
	(Percent)			
Crop				
Potatoes	22.04	.238	1.08	$/cwt.
Corn	15.72[2]	.099	.63	$/bu.
Tame Hay	11.95	.983	8.23	$/ton
Livestock				
Hogs	20.11	1.510	7.51	$/Head
Sheep	10.13	.269	2.66	$/Head
Mules	8.40	5.464	65.07	$/Head
Milk Cows	6.49	2.192	33.79	$/Head
All Cattle	5.92	1.268	21.41	$/Head
Horses	5.85[a]	2.199	37.60	$/Head

[1]All estimates are based on 28 observations unless otherwise noted, 1882–1909.
[2]This estimate is based on 18 observations, 1892–1909.

Selected Crops and Livestock, Wyoming: Ranking By
Yield and Number on Farms Random Variability Coefficients[1]

	Random Variability Coefficient	Standard Deviation	Mean	Unit
	(Percent)			
Crop				
Barley	22.32[2]	4.897	21.94	Bu./acre
Potatoes	16.34	10.306	63.07	Cwt./acre
Corn	10.33[3]	1.828	17.70	Bu./acre
Wheat	9.75[4]	1.751	17.97	Bu./acre
Tame Hay	8.38	.132	1.57	Ton/acre
Oats	8.35[3]	2.484	29.76	Bu./acre
Livestock				
Mules	9.91[a]	.113	1.1	1000's of Head
Hogs	5.66	1.025	18.1	1000's of Head
Sheep	3.90	101.551	2600.9	1000's of Head
All Cattle	3.82	27.703	724.9	1000's of Head
Milk Cows	2.72	.352	12.9	1000's of Head
Horses	.25	.263	107.3	1000's of Head

[1]All estimates are based on 28 observations unless otherwise noted, 1882-1909.
[2]This estimate is based on 11 observations, 1899-1909.
[3]These estimates are based on 21 observations, 1889-1909.
[4]This estimate is based on 20 observations, 1890-1909.

TABLE A120

Selected Crops, Wyoming: Ranking By
Gross Income Random Variability Coefficients

Crop	Random Variability Coefficient	Standard Deviation	Mean	Unit
	(Percent)			
Barley	25.30[1]	3.630	14.35	$/acre
Potatoes	23.39[2]	15.781	67.47	$/acre
Tame Hay	23.33[2]	2.981	12.78	$/acre
Corn	20.85[3]	2.328	11.16	$/acre
Oats	17.28[3]	2.301	13.32	$/acre
Wheat	16.68[4]	2.214	13.27	$/acre

[1]This estimate is based on 11 observations, 1899-1909.
[2]These estimates are based on 28 observations, 1882-1909.
[3]These estimates are based on 18 observations, 1892-1909.
[4]This estimate is based on 19 observations, 1891-1909.

Selected Crops and Livestock, Colorado: Ranking
By Price Random Variability Coefficients[1]

	Random Variability Coefficient	Standard Deviation	Mean	Unit
	(Percent)			
Crop				
Potatoes	33.82	.345	1.02	$/cwt.
Wheat	23.21	.174	.75	$/bu.
Corn	14.38	.085	.59	$/bu.
Oats	12.90	.059	.46	$/bu.
Barley	11.76	.074	.63	$/bu.
Rye	11.73	.075	.64	$/bu.
Tame Hay	5.26	.510	9.70	$/ton
Livestock				
Hogs	10.79	.765	7.09	$/Head
All Cattle	5.91	1.257	21.28	$/Head
Sheep	5.43	.131	2.42	$/Head
Horses	4.80	2.324	48.42	$/Head
Mules	4.67[a]	3.382	72.41	$/Head
Milk Cows	2.97	.978	32.97	$/Head

[1]All estimates are based on 30 observations, 1880-1909.

TABLE A122

Selected Crops and Livestock, Colorado: Ranking By
Yield and Number on Farms Random Variability Coefficients[1]

	Random Variability Coefficient	Standard Deviation	Mean	Unit
	(Percent)			
Crop				
Rye	23.59	2.819	11.95	Bu./acre
Oats	15.26	4.575	29.99	Bu./acre
Wheat	15.09	2.970	19.68	Bu./acre
Potatoes	13.21	7.256	54.93	Cwt./acre
Tame Hay	11.53	.223	1.94	Ton/acre
Corn	10.07	1.843	18.29	Bu./acre
Barley	8.04	2.111	26.27	Bu./acre
Livestock				
Mules	3.92	.289	7.4	1000's of Head
Sheep	1.85	22.698	1230.0	1000's of Head
Milk Cows	1.80	1.261	70.0	1000's of Head
Hogs	1.75	1.617	92.2	1000's of Head
All Cattle	1.12	13.263	1181.3	1000's of Head
Horses	.31	.627	200.4	1000's of Head

[1]All estimates are based on 30 observations, 1880-1909.

Selected Crops, Colorado: Ranking By
Gross Income Random Variability Coefficients[1]

Crop	Random Variability Coefficient	Standard Deviation	Mean	Unit
	(Percent)			
Rye	32.61	2.478	7.60	$/acre
Potatoes	30.99	16.904	54.54	$/acre
Wheat	26.66	3.922	14.71	$/acre
Barley	13.80	2.287	16.57	$/acre
Oats	13.55	1.867	13.78	$/acre
Tame Hay	11.92	2.194	18.40	$/acre
Corn	11.52	1.243	10.79	$/acre

[1] All estimates are based on 30 observations, 1880-1909.

TABLE A124

Selected Crops and Livestock, New Mexico: Ranking
By Price Random Variability Coefficients[1]

	Random Variability Coefficient	Standard Deviation	Mean	Unit
	(Percent)			
Crop				
Potatoes	27.00[2]	.357	1.32	$/cwt.
Tame Hay	23.68	2.507	10.58	$/ton
Wheat	21.78	.188	.86	$/bu.
Barley	18.91	.131	.70	$/bu.
Corn	6.52	.047	.72	$/bu.
Oats	4.60	.024	.53	$/bu.
Livestock				
Sheep	13.75	.251	1.83	$/Head
Mules	12.19	5.603	45.98	$/Head
Hogs	11.45	.713	6.22	$/Head
Horses	7.85	2.260	28.77	$/Head
All Cattle	5.45	.836	15.33	$/Head
Milk Cows	4.80	1.349	28.12	$/Head

[1] All estimates are based on 28 observations unless otherwise noted, 1882-1909.
[2] This estimate is based on 25 observations, 1885-1909.

Selected Crops and Livestock, New Mexico: Ranking By [1]
Yield and Number on Farms Random Variability Coefficients

	Random Variability Coefficient	Standard Deviation	Mean	Unit
	(Percent)			
Crop				
Barley	22.92	5.401	23.56	Bu./acre
Oats	19.22	4.522	23.53	Bu./acre
Potatoes	18.77[2]	7.367	39.24	Cwt./acre
Wheat	15.36	2.716	17.68	Bu./acre
Tame Hay	9.00	.197	2.19	Ton/acre
Corn	8.13	1.690	20.80	Bu./acre
Livestock				
Hogs	4.48	1.230	27.5	1000's of Head
Sheep	2.69	93.248	3461.8	1000's of Head
Milk Cows	1.82	.320	17.6	1000's of Head
Mules	1.82	.122	6.7	1000's of Head
All Cattle	1.56	16.385	1050.4	1000's of Head
Horses	.88	.963	109.9	1000's of Head

[1]All estimates are based on 28 observations unless otherwise noted, 1882-1909.
[2]This estimate is based on 25 observations, 1885-1909.

TABLE A126

Selected Crops, New Mexico: Ranking By [1]
Gross Income Random Variability Coefficients

Crop	Random Variability Coefficient	Standard Deviation	Mean	Unit
	(Percent)			
Potatoes	35.31[2]	18.557	52.55	$/acre
Barley	30.26	4.934	16.30	$/acre
Tame Hay	23.56	5.441	23.10	$/acre
Oats	18.08	2.192	12.12	$/acre
Wheat	17.27	2.592	15.01	$/acre
Corn	9.67	1.449	14.98	$/acre

[1]All estimates are based on 28 observations unless otherwise noted, 1882-1909.
[2]This estimate is based on 25 observations, 1885-1909.

Selected Crops and Livestock, Arizona: Ranking By Price Random Variability Coefficients[1]

	Random Variability Coefficient	Standard Deviation	Mean	Unit
	(Percent)			
Crop				
Tame Hay	24.34	2.744	11.27	$/ton
Wheat	21.92	.203	.93	$/bu.
Barley	21.36[c,2]	.146	.68	$/bu.
Potatoes	9.46	.193	2.03	$/cwt.
Livestock				
Hogs	23.03	1.450	6.29	$/Head
Milk Cows	12.51	3.805	30.41	$/Head
Sheep	10.74[a]	.241	2.24	$/Head
Mules	9.75	5.611	57.57	$/Head
Horses	8.83	3.348	37.93	$/Head
All Cattle	2.85	.472	16.58	$/Head

[1]All estimates are based on 28 observations unless otherwise noted, 1882–1909.
[2]This estimate is based on 14 observations, 1882–1895.
[3]This estimate is based on 11 observations, 1899–1909.

TABLE A128

Selected Crops and Livestock, Arizona: Ranking By Yield and Number on Farms Random Variability Coefficients[1]

	Random Variability Coefficient	Standard Deviation	Mean	Unit
	(Percent)			
Crop				
Wheat	12.04	2.081	17.28	Bu./acre
Potatoes	11.20[c,2]	4.338	38.73	Cwt./acre
Tame Hay	9.02	.225	2.50	Ton/acre
Oats	8.21[2]	2.590	31.56	Bu./acre
Corn	7.25	1.435	19.79	Bu./acre
Barley	5.14	1.381	26.88	Bu./acre
Livestock				
Mules	7.73	.226	2.9	1000's of Head
Milk Cows	2.45	.376	15.3	1000's of Head
Sheep	2.03	15.763	774.9	1000's of Head
Hogs	1.74	.200	11.5	1000's of Head
Horses	.78	.583	74.8	1000's of Head
All Cattle	.76	6.733	884.4	1000's of Head

[1]All estimates are based on 28 observations unless otherwise noted, 1882–1909.
[2]These estimates are based on 11 observations, 1899–1909.

Selected Crops, Arizona: Ranking By
Gross Income Random Variability Coefficients

Crop	Random Variability Coefficient	Standard Deviation	Mean	Unit
	(Percent)			
Tame Hay	24.02[1]	6.723	27.99	$/acre
Barley	21.54[c,2]	3.775	17.52	$/acre
Wheat	15.46[1]	2.486	16.08	$/acre
Potatoes	14.18[c,3]	11.229	79.18	$/acre

[1]These estimates are based on 28 observations, 1882–1909.
[2]This estimate is based on 14 observations, 1882–1895.
[3]This estimate is based on 11 observations, 1899–1909.

TABLE A130

Selected Crops and Livestock, Utah: Ranking
By Price Random Variability Coefficients[1]

	Random Variability Coefficient	Standard Deviation	Mean	Unit
	(Percent)			
Crop				
Potatoes	31.59	.244	.77	$/cwt.
Rye	18.28	.108	.59	$/bu.
Tame Hay	11.39	.763	6.70	$/ton
Oats	11.04	.048	.43	$/bu.
Corn	10.30	.069	.67	$/bu.
Barley	8.61	.047	.55	$/bu.
Wheat	7.99	.056	.70	$/bu.
Livestock				
Hogs	10.90	.857	7.86	$/Head
Mules	7.86[a]	3.771	47.95	$/Head
Milk Cows	7.50	2.036	27.15	$/Head
All Cattle	6.95	1.329	19.14	$/Head
Sheep	5.39	.131	2.43	$/Head
Horses	4.68	1.789	38.21	$/Head

[1]All estimates are based on 28 observations, 1882–1909.

Selected Crops and Livestock, Utah: Ranking By
Yield and Number on Farms Random Variability Coefficients[1]

	Random Variability Coefficient	Standard Deviation	Mean	Unit
	(Percent)			
Crop				
Rye	16.80	1.980	11.79	Bu./acre
Wheat	13.74	2.900	21.11	Bu./acre
Potatoes	12.79	10.030	78.43	Cwt./acre
Corn	10.50	2.188	20.85	Bu./acre
Barley	8.66	2.704	31.21	Bu./acre
Tame Hay	6.62	.161	2.42	Ton/acre
Oats	4.87	1.627	33.43	Bu./acre
Livestock				
Mules	7.21	.113	1.6	1000's of Head
Sheep	3.65	76.220	2086.6	1000's of Head
Hogs	1.37	.704	51.5	1000's of Head
All Cattle	1.23	4.538	367.4	1000's of Head
Milk Cows	.89	.460	51.9	1000's of Head
Horses	.33	.323	98.4	1000's of Head

[1]All estimates are based on 28 observations, 1882-1909.

TABLE A132

Selected Crops, Utah: Ranking By
Gross Income Random Variability Coefficients[1]

Crop	Random Variability Coefficient	Standard Deviation	Mean	Unit
	(Percent)			
Rye	28.85	2.019	7.00	$/acre
Wheat	23.59	3.497	14.83	$/acre
Potatoes	21.82	13.072	59.90	$/acre
Tame Hay	20.10	3.244	16.14	$/acre
Corn	13.71	1.927	14.06	$/acre
Barley	12.35	2.097	16.98	$/acre
Oats	10.34	1.496	14.46	$/acre

[1]All estimates are based on 28 observations, 1882-1909.

Selected Crops and Livestock, Nevada: Ranking
By Price Random Variability Coefficients

	Random Variability Coefficient	Standard Deviation	Mean	Unit
	(Percent)			
Crop				
Wheat	19.90[c,1]	.190	.96	$/bu.
Potatoes	16.95[2]	.209	1.23	$/cwt.
Oats	15.09[c,1]	.098	.65	$/bu.
Tame Hay	10.01[c,1]	1.957	19.51	$/ton
Barley	8.14[3]	.069	.85	$/bu.
Livestock[4]				
Hogs	14.20	1.042	7.34	$/Head
Horses	13.32[a]	6.207	46.59	$/Head
Mules	12.03	7.741	64.37	$/Head
Sheep	10.10	.263	2.60	$/Head
Milk Cows	8.38	2.876	34.30	$/Head
All Cattle	7.56	1.539	20.36	$/Head

[1]These estimates are based on 7 observations, 1870-1876.
[2]This estimate is based on 40 observations, 1870-1909.
[3]This estimate is based on 31 observations, 1879-1909.
[4]All livestock estimates are based on 40 observations, 1870-1909.

Selected Crops and Livestock, Nevada: Ranking By
Yield and Number on Farms Random Variability Coefficients

	Random Variability Coefficient	Standard Deviation	Mean	Unit
	(Percent)			
Crop[1]				
Wheat	13.72	3.607	26.30	Bu./acre
Tame Hay	6.36	.129	2.03	Ton/acre
Oats	6.28[c,2]	2.341	37.30	Bu./acre
Potatoes	6.01[3]	5.920	98.58	Cwt./acre
Barley	5.75[3]	1.747	30.39	Bu./acre
Livestock[3]				
Mules	7.86	.138	1.8	1000's of Head
Milk Cows	2.40	.248	10.3	1000's of Head
Hogs	1.72	.201	11.7	1000's of Head
Sheep	1.14	5.529	484.8	1000's of Head
All Cattle	1.09	3.067	282.2	1000's of Head
Horses	.91	3.807	50.6	1000's of Head

[1]All crop estimates are based on 31 observations unless otherwise noted, 1879-1909.

[2]This estimate is based on 11 observations, 1899-1909.
[3]These estimates are all based on 40 observations, 1870-1909.

TABLE A135

Selected Crops, Nevada: Ranking By
Gross Income Random Variability Coefficients

Crop	Random Variability Coefficient	Standard Deviation	Mean	Unit
	(Percent)			
Wheat	22.16[1]	5.220	23.56	$/acre
Potatoes	17.59[2]	21.365	121.46	$/acre
Tame Hay	16.18[1]	2.937	18.51	$/acre
Oats	11.07[c,3]	2.685	24.25	$/acre
Barley	7.23[a,4]	1.850	25.58	$/acre

[1]These estimates are based on 30 observations, 1880-1909.
[2]This estimate is based on 31 observations, 1879-1909.
[3]This estimate is based on 9 observations, 1901-1909.
[4]This estimate is based on 40 observations, 1870-1909.

Selected Crops and Livestock, Washington: Ranking
By Price Random Variability Coefficients[1]

	Random Variability Coefficient	Standard Deviation	Mean	Unit
	(Percent)			
Crop				
Potatoes	27.67	.214	.78	$/cwt.
Barley	25.14	.128	.51	$/bu.
Rye	16.47	.114	.69	$/bu.
Oats	15.52	.063	.41	$/bu.
Wheat	12.74	.084	.66	$/bu.
Tame Hay	7.41	.738	9.95	$/ton
Livestock				
Hogs	10.42	.676	6.49	$/Head
Mules	7.58	5.590	73.72	$/Head
Horses	5.95	3.559	59.82	$/Head
All Cattle	3.82	.904	23.66	$/Head
Sheep	2.90	.076	2.63	$/Head
Milk Cows	1.46	.485	33.28	$/Head

[1]All estimates are based on 28 observations, 1882–1909.

TABLE A137

Selected Crops and Livestock, Washington: Ranking By
Yield and Number on Farms Random Variability Coefficients[1]

	Random Variability Coefficient	Standard Deviation	Mean	Unit
	(Percent)			
Crop				
Rye	18.78	2.590	13.80	Bu./acre
Barley	13.33	3.869	29.03	Bu./acre
Corn	13.15	2.806	21.34	Bu./acre
Wheat	12.53[a]	2.358	18.84	Bu./acre
Oats	10.08	4.404	43.68	Bu./acre
Potatoes	9.93	7.645	77.00	Cwt./acre
Tame Hay	5.21	.085	1.63	Ton/acre
Livestock				
Mules	9.21	.253	2.8	1000's of Head
Hogs	5.72	9.447	165.1	1000's of Head
Sheep	2.50	12.959	519.4	1000's of Head
Milk Cows	2.31	2.159	93.5	1000's of Head
All Cattle	2.09	8.726	418.5	1000's of Head
Horses	1.38	2.640	191.8	1000's of Head

[1]All estimates are based on 28 observations, 1882–1909.

Selected Crops, Washington: Ranking By
Gross Income Random Variability Coefficients[1]

Crop	Random Variability Coefficient	Standard Deviation	Mean	Unit
	(Percent)			
Barley	28.26	4.198	14.86	$/acre
Wheat	23.49[a]	2.928	12.46	$/acre
Rye	17.57	1.657	9.43	$/acre
Potatoes	16.40	9.706	59.19	$/acre
Tame Hay	14.22	2.318	16.30	$/acre
Oats	13.71[a]	2.431	17.74	$/acre

[1]All estimates are based on 28 observations, 1882-1909.

TABLE A139

Selected Crops and Livestock, Oregon: Ranking
By Price Random Variability Coefficients[1]

	Random Variability Coefficient	Standard Deviation	Mean	Unit
	(Percent)			
Crop				
Potatoes	28.11	.264	.94	$/cwt.
Buckwheat	19.71[2]	.136	.69	$/bu.
Tame Hay	18.67	1.869	10.01	$/ton
Rye	17.52[3]	.131	.75	$/bu.
Oats	14.83[a]	.066	.45	$/bu.
Wheat	14.74	.113	.77	$/bu.
Barley	11.97	.068	.57	$/bu.
Corn	10.34[a]	.077	.74	$/bu.
Livestock				
Hogs	13.53	.606	4.48	$/Head
Sheep	8.19	.180	2.20	$/Head
Horses	6.38	3.244	50.88	$/Head
Mules	5.73	3.309	57.78	$/Head
All Cattle	4.79	.943	19.67	$/Head
Milk Cows	4.55	1.287	28.32	$/Head

[1]All estimates are based on 41 observations unless otherwise noted, 1869-1909.
[2]This estimate is based on 20 observations, 1881-1900.
[2]This estimate is based on 33 observations, 1876-1909.

Selected Crops and Livestock, Oregon: Ranking By
Yield and Number on Farms Random Variability Coefficients[1]

	Random Variability Coefficient	Standard Deviation	Mean	Unit
	(Percent)			
Crop				
Rye	18.15[2]	2.228	12.28	Bu./acre
Potatoes	16.31	10.568	64.78	Cwt./acre
Oats	13.66	3.907	28.61	Bu./acre
Barley	11.40	3.218	28.22	Bu./acre
Wheat	9.01	1.673	18.58	Bu./acre
Corn	7.79	1.880	24.12	Bu./acre
Tame Hay	7.08	.103	1.45	Ton/acre
Livestock				
Mules	5.13	.261	5.1	1000's of Head
Hogs	2.97	5.981	201.3	1000's of Head
All Cattle	2.82	17.378	616.9	1000's of Head
Sheep	1.81	32.232	1784.0	1000's of Head
Milk Cows	1.17	1.056	90.5	1000's of Head
Horses	.53	1.052	198.5	1000's of Head

[1]All estimates are based on 41 observations unless otherwise noted, 1869–1909.
[2]This estimate is based on 32 observations, 1878–1909.

TABLE A141

Selected Crops, Oregon: Ranking By
Gross Income Random Variability Coefficients[1]

Crop	Random Variability Coefficient	Standard Deviation	Mean	Unit
	(Percent)			
Potatoes	20.77	12.978	60.57	$/acre
Rye	18.67[a,2]	1.702	9.12	$/acre
Wheat	17.32	2.484	14.35	$/acre
Oats	17.19	2.208	12.85	$/acre
Tame Hay	16.21	2.330	14.37	$/acre
Barley	15.28	2.471	16.17	$/acre
Corn	12.27	2.198	17.92	$/acre

[1]All estimates are based on 41 observations unless otherwise noted, 1869–1909.
[2]This estimate is based on 32 observations, 1878–1909.

Selected Crops and Livestock, California: Ranking
By Price Random Variability Coefficients[1]

	Random Variability Coefficient	Standard Deviation	Mean	Unit
	(Percent)			
Crop				
Potatoes	19.10[2]	.234	1.23	$/cwt.
Barley	17.53	.119	.68	$/bu.
Sweet Potatoes	16.74[3]	.223	1.33	$/cwt.
Tame Hay	16.33	1.989	12.18	$/ton
Rye	14.02	.123	.88	$/bu.
Wheat	12.99	.121	.93	$/bu.
Corn	12.32	.098	.79	$/bu.
Oats	9.07	.054	.60	$/bu.
Livestock				
Hogs	12.37	.731	5.91	$/Head
Sheep	7.46[a]	.178	2.39	$/Head
Mules	5.76[a]	4.251	73.76	$/Head
All Cattle	5.61	1.357	24.17	$/Head
Horses	4.20	2.364	56.29	$/Head
Milk Cows	3.84	1.360	35.44	$/Head

[1]All estimates are based on 42 observations unless otherwise noted, 1868–1909.
[2]This estimate is based on 35 observations, 1868–1902.
[3]This estimate is based on 21 observations, 1889–1909.

Selected Crops and Livestock, California: Ranking By
Yield and Number on Farms Random Variability Coefficients[1]

	Random Variability Coefficient	Standard Deviation	Mean	Unit
	(Percent)			
Crop				
Rye	23.22	2.098	9.00	Bu./acre
Barley	17.94	3.730	20.79	Bu./acre
Wheat	16.01	2.218	13.85	Bu./acre
Oats	12.71	3.541	27.86	Bu./acre
Sweet Potatoes	11.45[2]	8.835	77.19	Cwt./acre
Potatoes	10.81	7.023	64.95	Cwt./acre
Corn	10.47	3.099	29.59	Bu./acre
Tame Hay	7.88	.113	1.44	Ton/acre
Livestock				
Sheep	3.33	137.712	4141.0	1000's of Head
All Cattle	1.22	16.575	1359.4	1000's of Head
Hogs	1.19	8.306	699.3	1000's of Head
Milk Cows	.94	2.456	260.8	1000's of Head
Mules	.81	.418	51.5	1000's of Head
Horses	.50	1.753	348.6	1000's of Head

[1]All estimates are based on 42 observations unless otherwise noted, 1868–1909.
[2]This estimate is based on 21 observations, 1889–1909.

TABLE A144

Selected Crops, California: Ranking By
Gross Income Random Variability Coefficients[1]

Crop	Random Variability Coefficient	Standard Deviation	Mean	Unit
	(Percent)			
Rye	22.40	1.801	8.04	$/acre
Barley	17.88	2.479	13.86	$/acre
Wheat	17.14	2.221	12.96	$/acre
Potatoes	16.21[2]	11.756	72.51	$/acre
Oats	15.70[a]	2.579	16.43	$/acre
Corn	13.85	3.217	23.23	$/acre
Tame Hay	13.69	2.380	17.39	$/acre
Sweet Potatoes	13.34[3]	13.562	101.67	$/acre

[1]All estimates are based on 42 observations unless otherwise noted, 1868–1909.
[2]This estimate is based on 35 observations, 1868–1902.
[3]This estimate is based on 21 observations, 1889–1909.

APPENDIX B:

DISCUSSION OF SUBJECTIVE RISK COEFFICIENT CALCULATIONS

(TABLES B1 THROUGH B9 INCLUDED)

This appendix provides the estimates used in the calculations of
the subjective risk coefficients (b's). Because only the portion of total
variability that is unpredictable from the standpoint of the individual
farmer is relevant to behavior under uncertainty, the variability measures—
variances and covariances—were estimated using the variate difference
method. All variables concerning crop returns are for returns per acre
and measured in dollars. The acreage variables are listed as a fraction
of one and are actually the value for the year following each eight-year
interval estimated.

The following equation was used to compute our estimates of b:

$$b = \frac{\mu_x - \mu_y}{2[a\sigma_x^2 - (1-a)\sigma_y^2 + (1-2a)\ \sigma_{xy}]}.$$

Keeping this equation in mind, an inspection of Tables B1 through B9 leads
to a number of interesting observations.

Because the expected return from cotton at each interval for all
states was larger than the expected return of corn, the numerator of the
above equation was always negative. As long as there is a negative denominator, our estimate of "b" always will be positive. Thus, it will imply
risk averse behavior on the part of Southern farmers. Because the random
variance of cotton at each interval for all states was larger than the
random variance of corn, the denominator will tend to be negative, unless
the covariance term outweighs the effect of cotton variance. However, the
covariance term is never an important determinant of the sign of the denominator. This is true because the covariance is usually quite small (although
not always), and only a fraction of it (1-2a) is used in the calculation.
Also, the proportion of total acreage in corn ("a") will determine the
weights attached to the two variances.

Using the above information, it becomes obvious that the large differential in expected returns between the two crops allowed the Southern farmers to choose a particular crop-mix with more acreage in cotton than corn and not be gambling. In fact, with the particular means and variances faced by the farmers, an exceptionally large proportion of acreage to corn could have implied gambling behavior. However, this usually would not be the case, because of the large differential in variances between the two crops. Also, if the farmers had chosen an exceptionally large proportion of acreage to cotton, given these same means and variances, the denominator could have been quite large and the resulting estimate insignificantly different than zero. Therefore, this situation could imply an approximation to risk preferring behavior. Yet it would actually show risk neutral behavior.

The implication from the preceding is quite clear. The crop-mix chosen by postbellum farmers indicates, at least according to our model, risk averse behavior.

TABLE B1

Estimates of Variables Necessary for Subjective Risk Coefficient
Calculations, North Carolina: Computed at
Eight-Year Intervals

Description of Variable	Value for 1869–76	Value for 1877–84	Value for 1885–92	Value for 1893–00	Value for 1901–08
Mean Return Per Acre, Corn (μ_x)	8.75	6.94	6.52	5.92	9.51
Mean Return Per Acre, Cotton (μ_x)	23.14	17.50	14.33	15.59	21.21
Random Variance, Corn Returns (σ_x^2)	1.306	.547	.717	.584	.381
Random Variance, Cotton Returns (σ_x^2)	11.528	13.885	5.166	6.672	7.760
Random Covariance, Returns (σ_{xy})	− .066	1.985	1.157	− .240	− .588
Percent of Acres Harvested, Corn (a)	.752	.691	.691	.700	.670
Percent of Acres Harvested, Cotton (1−a)	.248	.309	.309	.300	.330
Covariance Multiple (1−2a)	− .503	− .381	− .381	− .400	− .340

TABLE B2

Estimates of Variables Necessary for Subjective Risk Coefficient
Calculations, South Carolina: Computed at
Eight-Year Intervals

Description of Variable	Value for 1869–76	Value for 1877–84	Value for 1885–92	Value for 1893–00	Value for 1901–08
Mean Return per Acre, Corn (μ_x)	9.77	6.71	6.06	5.34	9.10
Mean Return per Acre, Cotton (μ_y)	24.84	15.53	12.46	14.10	20.08
Random Variance, Corn Returns (σ_x^2)	1.380	.792	.851	1.045	1.459
Random Variance, Cotton Returns (σ_y^2)	31.866	7.961	1.758	6.450	17.007
Random Covariance, Returns (σ_{xy})	3.318	1.847	.697	− 1.612	1.782
Percent of Acres Harvested, Corn (a)	.546	.443	.446	.443	.410
Percent of Acres Harvested, Cotton (1−a)	.454	.557	.554	.557	.590
Covariance Multiple (1−2a)	− .092	.114	.108	.114	.180

TABLE B3

Estimates of Variables Necessary for Subjective Risk Coefficient
Calculations, Georgia: Computed at
Eight-Year Intervals

Description of Variable	Value for 1869-76	Value for 1877-84	Value for 1885-92	Value for 1893-00	Value for 1901-08
Mean Return Per Acre, Corn (μ_x)	8.92	6.76	6.65	5.42	8.19
Mean Return Per Acre, Cotton (μ_y)	21.88	12.32	12.30	11.89	17.76
Random Variance, Corn Returns (σ_y^2)	1.485	.814	.580	.309	.573
Random Covariance, Cotton Returns (σ_y^2)	17.861	4.649	2.037	2.410	7.388
Random Covariance, Returns (σ_{xy})	.441	.135	- .174	- .206	.488
Percent of Acres Harvested, Corn (a)	.540	.444	.496	.481	.428
Percent of Acres Harvested, Cotton (1-a)	.460	.556	.504	.519	.572
Covariance Multiple (1-2a)	- .080	.112	.008	.038	.144

TABLE B4

Estimates of Variables Necessary for Subjective Risk Coefficient
Calculations, Florida: Computed at
Eight-Year Intervals

Description of Variable	Value for 1869–76	Value for 1877–84	Value for 1885–92	Value for 1893–00	Value for 1901–08
Mean Return Per Acre, Corn (μ_x)	11.08	7.61	6.47	5.29	7.25
Mean Return Per Acre, Cotton (μ_x)	17.00	12.77	12.04	8.32	13.87
Random Variance, Corn Returns (σ_x^2)	1.914	.227	.747	.250	.522
Random Variance, Cotton Returns (σ_y^2)	3.255	19.439	18.290	6.537	4.693
Random Covariance, Returns (σ_{xy})	2.713	– .804	– .192	– .484	.602
Percent of Acres Harvested, Corn (a)	.612	.629	.709	.713	.700
Percent of Acres Harvested, Cotton (1–a)	.388	.371	.291	.287	.300
Covariance Multiple (1–2a)	– .224	– .258	– .418	– .426	– .400

TABLE B5

Estimates of Variables Necessary for Subjective Risk Coefficient
Calculations, Alabama: Computed at
Eight-Year Intervals

Description of Variable	Value for 1869–76	Value for 1877–84	Value for 1885–92	Value for 1893–00	Value for 1901–08
Mean Return Per Acre, Corn (μ_x)	10.54	8.30	7.41	6.29	9.06
Mean Return Per Acre, Cotton (μ_y)	21.77	12.43	11.85	10.19	15.47
Random Variance, Corn Returns (σ_x^2)	1.617	.997	.522	.518	1.379
Random Variance, Cotton Returns (σ_y^2)	7.776	5.787	3.094	2.054	3.413
Random Covariance, Returns (σ_{xy})	.537	– .112	– .091	– .552	1.111
Percent of Acres Harvested, Corn (a)	.501	.457	.463	.424	.428
Percent of Acres Harvested, Cotton (1-a)	.499	.543	.537	.576	.572
Covariance Multiple (1-2a)	– .002	.086	.074	.152	.144

TABLE B6

Estimates of Variables Necessary for Subjective Risk Coefficient
Calculations, Mississippi: Computed at
Eight-Year Intervals

Description of Variable	Value for 1869–76	Value for 1877–84	Value for 1885–92	Value for 1893–00	Value for 1901–08
Mean Return Per Acre, Corn (μ_x)	12.18	8.85	8.01	6.99	10.24
Mean Return Per Acre, Cotton (μ_y)	27.85	17.42	15.26	13.57	18.74
Random Variance, Corn Returns (σ_x^2)	3.664	.459	.238	.522	1.276
Random Variance, Cotton Returns (σ_y^2)	15.810	6.951	2.331	4.304	2.197
Random Covariance, Returns (σ_{xy})	3.329	− .652	.306	− .821	.514
Percent of Acres Harvested, Corn (a)	.428	.392	.416	.400	.400
Percent of Acres Harvested, Cotton (1−a)	.572	.608	.584	.600	.600
Covariance Multiple (1−2a)	.144	.216	.168	.200	.200

TABLE B7

Estimates of Variables Necessary for Subjective Risk Coefficient
Calculations, Arkansas: Computed at
Eight-Year Intervals

Description of Variable	Value for 1869–76	Value for 1877–84	Value for 1885–92	Value for 1893–00	Value for 1901–08
Mean Return Per Acre, Corn (μ_x)	13.05	10.32	9.51	7.04	10.28
Mean Return Per Acre, Cotton (μ_y)	31.60	22.06	17.20	14.43	19.04
Random Variance, Corn Returns (σ_x^2)	2.850	2.547	.540	.897	1.647
Random Variance, Cotton Returns (σ_y^2)	46.940	17.798	1.900	7.305	2.798
Random Covariance, Returns (σ_{xy})	− .111	−3.564	.604	− .010	1.841
Percent of Acres Harvested, Corn (a)	.551	.540	.566	.544	.540
Percent of Acres Harvested, Cotton (1–a)	.449	.460	.434	.456	.460
Covariance Multiple (1–2a)	− .102	− .080	− .132	− .088	− .080

TABLE B8

Estimates of Variables Necessary for Subjective Risk Coefficient
Calculations, Louisiana: Computed at
Eight-Year Intervals

Description of Variable	Value for 1869–76	Value for 1877–84	Value for 1885–92	Value for 1893–00	Value for 1901–08
Mean Return Per Acre, Corn (μ_x)	14.12	9.67	8.57	7.62	9.37
Mean Return Per Acre, Cotton (μ_y)	35.03	22.47	18.20	15.79	20.97
Random Variance, Corn Returns (σ_x^2)	2.234	1.000	.744	1.181	1.517
Random Variance, Cotton Returns (σ_y^2)	27.508	11.993	5.862	5.710	22.550
Random Covariance, Returns (σ_{xy})	4.024	.262	1.115	- .649	.557
Percent of Acres Harvested, Corn (a)	.414	.439	.443	.440	.630
Percent of Acres Harvested, Cotton (1–a)	.586	.561	.557	.560	.370
Covariance Multiple (1–2a)	.172	.122	.114	.120	- .260

TABLE B9

Estimates of Variables Necessary for Subjective Risk Coefficient
Calculations, Texas: Computed at
Eight-Year Intervals

Description of Variable	Value for 1869–76	Value for 1877–84	Value for 1885–92	Value for 1893–00	Value for 1901–08
Mean Return Per Acre, Corn (μ_x)	15.21	11.68	10.44	8.20	11.36
Mean Return Per Acre, Cotton (μ_y)	31.37	17.54	16.94	13.34	14.51
Random Variance, Corn Returns (σ_x^2)	8.430	.908	1.870	2.917	5.357
Random Variance, Cotton Returns (σ_y^2)	23.342	9.495	2.523	5.384	11.912
Random Covariance, Returns (σ_{xy})	−3.417	1.930	1.025	1.946	3.239
Percent of Acres Harvested, Corn (a)	.562	.518	.437	.372	.343
Percent of Acres Harvested, Cotton (1−a)	.438	.412	.563	.628	.657
Covariance Multiple (1−2a)	− .124	− .036	.126	.256	.314

APPENDIX C:

DISCUSSION OF RISK MEASURES AND AGRARIAN DISCONTENT

(TABLES C1 THROUGH C24 INCLUDED)

This appendix contains the random variability coefficients (presented in Tables C1 through C12) which were used to explain the differences in the level of agrarian unrest across states, and the random variances (presented in Tables C13 through C24) which were used to explain differences in the temporal pattern of farmer discontent.

Tables C1 through C12. The data for these estimates were the revised statistics of the Department of Agriculture, the sources of which are listed in Appendix A. The computations were made by using the variate difference method of estimating a random variance. The state estimates are listed in the order in which the state is listed in the census regions; in addition, their ranking (according to the level of variability) is indicated by the number in parentheses preceding each estimate—for example, (1) refers to the estimate with the highest level of variability for that series and (2) refers to the next highest estimate. Those series for which data were not available (N.A.) are indicated as such in the tables. It is interesting to note that the price variability estimates for the Dakotas during the Alliance period (Table C7) were identical. The reason for this is quite simple. Price estimates collected by the U.S.D.A. for North and South Dakota actually were collected for the Dakota territories until after 1890. Thus, the two states' farm prices were identical during the Alliance period.

Tables C13 through C24. The data for these estimates were the revised statistics of the Department of Agriculture, the sources of which are listed in Appendix A. The variate difference method was used to compute the portion of total variance of each series that is random from the standpoint of an individual farmer. The state estimates are presented in the order in which they appear in their respective census regions. Those series for which data were not available (N.A.) are indicated as such in the tables. The price and income variances of cotton during the Granger period (1867-1874) actually were estimates for the period 1869-1874, because no cotton price series existed prior to 1869. In addition, this study used a national average price of cotton prior to 1876; thus, the price variabilities of cotton are identical for every state during the Granger period.

TABLE C1

Wheat, Oats, and Corn; Granger Period, 1867–1874: Ranking by Price
Random Variability Coefficients for Northern States

State	Wheat Random Variability Coefficient (Percent)	Corn Random Variability Coefficient (Percent)	Oats Random Variability Coefficient (Percent)
New York	(10) 15.68	(11) 10.47	(9) 12.81
Pennsylvania	(12) 10.42	(12) 6.78	(8) 13.27
Ohio	(9) 19.01	(7) 17.57	(11) 12.11
Indiana	(7) 19.89	(5) 21.44	(7) 13.56
Illinois	(5) 21.96	(3) 26.72	(3) 18.61
Michigan	(8) 19.76	(9) 14.67	(6) 14.91
Wisconsin	(3) 22.42	(8) 16.92	(10) 12.56
Minnesota	(1) 22.60	(6) 20.37	(4) 18.42
Iowa	(4) 22.40	(4) 23.87	(5) 17.87
Missouri	(11) 14.70	(10) 12.14	(12) 8.64
North Dakota	N.A.	N.A.	N.A.
South Dakota	N.A.	N.A.	N.A.
Nebraska	(2) 22.50	(2) 35.42	(2) 25.94
Kansas	(6) 21.86	(1) 50.92	(1) 29.15

TABLE C2

Wheat, Oats, and Corn; Granger Period, 1867–1874: Ranking by Yield
Random Variability Coefficients for Northern States

State	Wheat Random Variability Coefficient (Percent)	Corn Random Variability Coefficient (Percent)	Oats Random Variability Coefficient (Percent)
Ohio	(9) 11.01	(9) 9.52	(9) 7.89
Indiana	(3) 13.57	(4) 14.45	(8) 9.57
Illinois	(11) 7.92	(2) 24.63	(2) 14.40
Michigan	(8) 11.26	(6) 11.89	(12) 6.42
Wisconsin	(7) 11.45	(7) 11.14	(6) 11.70
Minnesota	(10) 10.46	(11) 8.98	(11) 7.52
Iowa	(12) 6.74	(8) 11.03	(5) 11.89
Missouri	(2) 14.78	(5) 13.07	(10) 7.86
North Dakota	N.A.	N.A.	N.A.
South Dakota	N.A.	N.A.	N.A.
Nebraska	(1) 18.80	(3) 19.70	(3) 14.31
Kansas	(6) 11.51	(1) 31.92	(1) 16.25
New York	(5) 12.40	(10) 9.04	(7) 11.67
Pennsylvania	(4) 13.26	(12) 6.71	(4) 14.01

TABLE C3

Wheat, Oats, and Corn; Granger Period, 1867-1874: Ranking by Income
Random Variability Coefficients for Northern States

State	Wheat Random Variability Coefficient (Percent)	Corn Random Variability Coefficient (Percent)	Oats Random Variability Coefficient (Percent)
New York	(11) 12.14	(10) 10.12	(11) 9.45
Pennsylvania	(6) 17.69	(11) 8.35	(12) 4.50
Ohio	(12) 10.90	(12) 6.94	(10) 10.69
Indiana	(9) 16.14	(9) 11.23	(8) 11.61
Illinois	(1) 23.51	(4) 19.00	(9) 11.07
Michigan	(10) 14.65	(8) 14.46	(5) 15.34
Wisconsin	(5) 21.05	(5) 18.87	(6) 13.53
Minnesota	(7) 17.62	(2) 20.85	(3) 17.24
Iowa	(4) 21.69	(3) 19.14	(1) 21.44
Missouri	(3) 21.81	(6) 16.32	(7) 11.93
North Dakota	N.A.	N.A.	N.A.
South Dakota	N.A.	N.A.	N.A.
Nebraska	(2) 22.47	(1) 23.54	(2) 18.49
Kansas	(8) 16.52	(7) 15.10	(4) 15.65

TABLE C4

Wheat, Oats, and Corn; Greenback Period, 1874-1881: Ranking by Price
Random Variability Coefficients for Northern States

State	Wheat Random Variability Coefficient (Percent)	Corn Random Variability Coefficient (Percent)	Oats Random Variability Coefficient (Percent)
New York	(12) 9.62	(11) 13.33	(12) 13.09
Pennsylvania	(11) 10.97	(10) 13.78	(8) 16.57
Ohio	(9) 12.43	(7) 16.52	(9) 16.51
Indiana	(8) 14.02	(6) 18.00	(10) 16.22
Illinois	(7) 14.82	(5) 23.83	(5) 23.38
Michigan	(10) 12.11	(12) 12.90	(11) 14.82
Wisconsin	(5) 16.22	(9) 15.38	(7) 16.69
Minnesota	(3) 19.86	(8) 15.78	(6) 18.63
Iowa	(1) 21.69	(4) 25.59	(4) 23.61
Missouri	(6) 16.16	(3) 39.59	(3) 24.71
North Dakota	N.A.	N.A.	N.A.
South Dakota	N.A.	N.A.	N.A.
Nebraska	(2) 20.86	(2) 52.20	(1) 32.07
Kansas	(4) 16.41	(1) 55.05	(2) 30.49

TABLE C5

Wheat, Oats, and Corn; Greenback Period, 1874–1881: Ranking by Yield
Random Variability Coefficients for Northern States

State	Wheat Random Variability Coefficient (Percent)	Corn Random Variability Coefficient (Percent)	Oats Random Variability Coefficient (Percent)
New York	(6) 18.49	(11) 7.35	(11) 6.34
Pennsylvania	(12) 6.74	(8) 10.38	(12) 6.04
Ohio	(9) 16.02	(10) 7.61	(10) 8.08
Indiana	(2) 21.78	(12) 7.21	(7) 11.76
Illinois	(10) 12.15	(4) 16.30	(1) 24.30
Michigan	(11) 10.59	(6) 10.86	(9) 9.52
Wisconsin	(4) 18.89	(5) 13.97	(8) 10.03
Minnesota	(3) 20.90	(7) 10.44	(5) 13.53
Iowa	(1) 27.46	(9) 9.00	(6) 12.93
Missouri	(8) 16.80	(1) 28.15	(3) 15.89
North Dakota	N.A.	N.A.	N.A.
South Dakota	N.A.	N.A.	N.A.
Nebraska	(5) 18.67	(3) 16.77	(2) 18.26
Kansas	(7) 17.29	(2) 26.46	(4) 15.45

TABLE C6

Wheat, Oats, and Corn; Greenback Period, 1874–1881: Ranking by Income
Random Variability Coefficients for Northern States

State	Wheat Random Variability Coefficient (Percent)	Corn Random Variability Coefficient (Percent)	Oats Random Variability Coefficient (Percent)
New York	(7) 21.26	(11) 11.26	(6) 16.97
Pennsylvania	(12) 9.84	(12) 10.79	(11) 14.51
Ohio	(6) 21.54	(7) 13.80	(9) 14.94
Indiana	(3) 26.98	(5) 15.70	(12) 8.33
Illinois	(8) 19.69	(9) 13.52	(3) 22.30
Michigan	(9) 15.26	(8) 13.73	(8) 15.35
Wisconsin	(4) 26.39	(2) 19.71	(10) 14.83
Minnesota	(5) 26.29	(4) 17.07	(7) 15.98
Iowa	(1) 31.90	(3) 19.16	(5) 21.08
Missouri	(10) 14.36	(10) 11.62	(4) 21.45
North Dakota	N.A.	N.A.	N.A.
South Dakota	N.A.	N.A.	N.A.
Nebraska	(2) 30.94	(1) 25.24	(1) 33.42
Kansas	(11) 14.15	(6) 14.46	(2) 22.32

TABLE C7

Wheat, Oats, and Corn; Alliance Period, 1883-1890: Ranking by Price
Random Variability Coefficients for Northern States

State	Wheat Random Variability Coefficient (Percent)	Corn Random Variability Coefficient (Percent)	Oats Random Variability Coefficient (Percent)
New York	(12) 12.12	(13) 10.20	(12) 13.82
Pennsylvania	(13) 11.64	(12) 10.89	(11) 14.12
Ohio	(11) 12.73	(8) 19.59	(9) 19.12
Indiana	(9) 14.37	(5) 21.21	(7) 20.88
Illinois	(8) 14.88	(6) 20.92	(4) 24.88
Michigan	(10) 12.79	(9) 17.53	(10) 18.05
Wisconsin	(5) 16.52	(10) 16.68	(6) 21.51
Minnesota	(4) 19.59	(11) 15.42	(8) 20.58
Iowa	(6) 16.26	(3) 25.65	(2) 27.73
Missouri	(7) 16.18	(4) 22.38	(5) 22.52
North Dakota	(2) 22.80	(7) 20.66	(13) 11.97
South Dakota	(2) 22.80	(7) 20.66	(13) 11.97
Nebraska	(1) 24.40	(1) 41.55	(1) 34.24
Kansas	(3) 21.47	(2) 36.17	(3) 27.70

TABLE C8

Wheat, Oats, and Corn; Alliance Period, 1883-1890: Ranking by Yield
Random Variability Coefficients for Northern States

State	Wheat Random Variability Coefficient (Percent)	Corn Random Variability Coefficient (Percent)	Oats Random Variability Coefficient (Percent)
New York	(10) 13.38	(14) 6.20	(12) 8.54
Pennsylvania	(7) 14.25	(13) 6.40	(5) 11.59
Ohio	(2) 21.12	(9) 14.07	(4) 11.63
Indiana	(5) 19.56	(6) 17.04	(7) 10.66
Illinois	(1) 21.93	(3) 19.77	(3) 13.12
Michigan	(13) 8.25	(10) 13.00	(14) 6.53
Wisconsin	(11) 11.33	(8) 14.98	(6) 11.23
Minnesota	(9) 13.89	(12) 11.87	(11) 8.91
Iowa	(14) 6.49	(7) 16.51	(8) 10.55
Missouri	(4) 20.64	(11) 12.54	(9) 9.85
North Dakota	(6) 14.75	(2) 20.56	(1) 17.91
South Dakota	(8) 14.18	(4) 19.08	(2) 17.64
Nebraska	(12) 11.09	(5) 19.01	(10) 9.36
Kansas	(3) 20.74	(1) 32.02	(13) 7.87

TABLE C9

Wheat, Oats, and Corn; Alliance Period, 1883-1890: Ranking by Income
Random Variability Coefficients for Northern States

State	Wheat Random Variability Coefficient (Percent)		Corn Random Variability Coefficient (Percent)		Oats Random Variability Coefficient (Percent)	
New York	(13)	7.22	(13)	8.97	(12)	9.34
Pennsylvania	(5)	18.93	(14)	8.30	(14)	7.72
Ohio	(10)	12.58	(12)	9.13	(8)	14.37
Indiana	(14)	6.85	(11)	11.26	(13)	9.11
Illinois	(7)	18.13	(5)	13.49	(11)	10.39
Michigan	(11)	12.16	(7)	12.98	(10)	11.75
Wisconsin	(6)	18.17	(3)	16.39	(5)	17.15
Minnesota	(9)	14.24	(6)	13.04	(7)	17.02
Iowa	(8)	16.09	(8)	12.05	(1)	23.24
Missouri	(12)	11.27	(2)	16.63	(9)	12.73
North Dakota	(2)	30.18	(1)	18.96	(2)	22.21
South Dakota	(4)	27.85	(10)	11.52	(6)	17.10
Nebraska	(3)	29.66	(4)	14.48	(3)	21.61
Kansas	(1)	32.46	(9)	11.84	(4)	18.85

TABLE C10

Wheat, Oats, Corn; Populist Period, 1890-1897: Ranking by Price
Random Variability Coefficients for Northern States

State	Wheat Random Variability Coefficient (Percent)		Corn Random Variability Coefficient (Percent)		Oats Random Variability Coefficient (Percent)	
New York	(13)	6.92	(14)	10.04	(10)	16.07
Pennsylvania	(14)	6.46	(13)	11.16	(14)	13.48
Ohio	(11)	13.25	(9)	15.16	(13)	14.00
Indiana	(7)	14.15	(10)	14.62	(12)	15.37
Illinois	(5)	14.61	(7)	18.29	(6)	19.59
Michigan	(10)	13.36	(11)	14.26	(11)	15.80
Wisconsin	(12)	13.24	(8)	15.86	(7)	18.01
Minnesota	(8)	13.84	(5)	21.29	(4)	21.65
Iowa	(9)	13.47	(4)	33.45	(3)	23.16
Missouri	(4)	14.71	(6)	20.64	(9)	17.02
North Dakota	(2)	16.37	(8)	17.74	(8)	17.54
South Dakota	(3)	16.29	(2)	30.65	(2)	25.98
Nebraska	(1)	16.57	(1)	42.95	(1)	36.20
Kansas	(6)	14.45	(3)	27.65	(5)	19.71

TABLE C11

Wheat, Oats, and Corn; Populist Period, 1890–1897: Ranking by Yield
Random Variability Coefficients for Northern States

State	Wheat Random Variability Coefficient (Percent)	Corn Random Variability Coefficient (Percent)	Oats Random Variability Coefficient (Percent
New York	(12) 11.20	(14) 5.97	(8) 14.55
Pennsylvania	(11) 13.30	(12) 12.76	(10) 14.35
Ohio	(8) 17.56	(7) 14.85	(12) 10.72
Indiana	(5) 21.17	(6) 16.10	(7) 15.17
Illinois	(4) 23.14	(5) 19.08	(5) 17.94
Michigan	(14) 8.67	(13) 11.16	(13) 10.25
Wisconsin	(13) 10.25	(9) 14.16	(14) 7.84
Minnesota	(6) 20.77	(10) 13.93	(11) 14.25
Iowa	(10) 13.73	(4) 26.96	(4) 24.99
Missouri	(9) 14.68	(11) 13.19	(6) 15.25
North Dakota	(1) 32.29	(8) 14.20	(9) 14.46
South Dakota	(3) 28.03	(1) 45.24	(1) 21.99
Nebraska	(7) 18.24	(2) 37.73	(3) 26.87
Kansas	(2) 28.21	(3) 32.55	(2) 27.99

TABLE C12

Wheat, Oats, and Corn; Populist Period, 1890–1897: Ranking by Income
Random Variability Coefficients for Northern States

State	Wheat Random Variability Coefficient (Percent)	Corn Random Variability Coefficient (Percent)	Oats Random Variability Coefficient (Percent)
New York	(14) 7.17	(13) 8.72	(13) 10.81
Pennsylvania	(13) 13.49	(7) 18.81	(14) 7.66
Ohio	(8) 21.04	(9) 13.35	(12) 13.01
Indiana	(5) 26.41	(3) 26.15	(3) 22.47
Illinois	(3) 30.28	(5) 19.35	(1) 26.42
Michigan	(12) 17.34	(14) 7.12	(10) 14.20
Wisconsin	(10) 19.21	(11) 9.40	(8) 18.46
Minnesota	(7) 24.58	(12) 9.04	(9) 15.68
Iowa	(11) 17.73	(8) 13.45	(5) 21.71
Missouri	(9) 20.46	(10) 9.77	(11) 13.70
North Dakota	(1) 38.70	(4) 22.00	(4) 21.77
South Dakota	(2) 37.19	(1) 31.59	(7) 20.74
Nebraska	(6) 26.36	(6) 18.96	(2) 26.08
Kansas	(4) 28.98	(2) 28.65	(6) 20.77

TABLE C13

Price Random Variance Estimates of Wheat; Selected States: Computed
for Granger, Greenback, Alliance, and Populist Periods

State	Random Variance 1867–1874	Random Variance 1874–1881	Random Variance 1883–1890	Random Variance 1890–1897
Indiana	.0650	.0211	.0133	.0099
Illinois	.0627	.0205	.0135	.0100
Wisconsin	.0528	.0235	.0156	.0080
Minnesota	.0390	.0279	.0188	.0077
Iowa	.0368	.0304	.0126	.0072
North Dakota	N.A.	N.A.	.0208	.0086
South Dakota	N.A.	N.A.	.0208	.0085
Nebraska	.0322	.0235	.0214	.0089
Kansas	.0622	.0184	.0200	.0072
North Carolina	.0322	.0181	.0095	.0057
South Carolina	.0364	.0169	.0110	.0093
Georgia	.0363	.0108	.0060	.0063
Alabama	.0340	.0214	.0035	.0050
Mississippi	.0253	.0141	.0062	.0089
Arkansas	.0273	.0356	.0061	.0058
Louisiana	N.A.	N.A.	N.A.	N.A.
Texas	.0546	.0314	.0095	.0068

TABLE C14

Price Random Variance Estimates of Corn; Selected States: Computed
for Granger, Greenback, Alliance, and Populist Periods

State	Random Variance 1867–1874	Random Variance 1874–1881	Random Variance 1883–1890	Random Variance 1890–1897
Indiana	.0102	.0051	.0057	.0023
Illinois	.0130	.0079	.0049	.0032
Wisconsin	.0089	.0046	.0040	.0032
Minnesota	.0133	.0039	.0029	.0047
Iowa	.0076	.0055	.0053	.0088
North Dakota	N.A.	N.A.	.0057	.0042
South Dakota	N.A.	N.A.	.0057	.0093
Nebraska	.0247	.0244	.0105	.0149
Kansas	.0849	.0406	.0109	.0074
North Carolina	.0095	.0087	.0006	.0012
South Carolina	.0194	.0142	.0046	.0041
Georgia	.0105	.0124	.0030	.0035
Alabama	.0089	.0179	.0033	.0039
Mississippi	.0127	.0170	.0044	.0031
Arkansas	.0102	.0-15	.0045	.0045
Louisiana	.0146	.0178	.0042	.0056
Texas	.0467	.0599	.0141	.0086

TABLE C15

Price Random Variance Estimates of Oats; Selected States: Computed
for Granger, Greenback, Alliance, and Populist Periods

State	Random Variance 1867–1874	Random Variance 1874–1881	Random Variance 1883–1890	Random Variance 1890–1897
Indiana	.0026	.0027	.0035	.0018
Illinois	.0041	.0048	.0043	.0025
Wisconsin	.0027	.0029	.0036	.0022
Minnesota	.0058	.0038	.0028	.0028
Iowa	.0030	.0034	.0043	.0028
North Dakota	N.A.	N.A.	.0010	.0020
South Dakota	N.A.	N.A.	.0010	.0037
Nebraska	.0088	.0073	.0054	.0069
Kansas	.0184	.0079	.0046	.0024
North Carolina	.0055	.0023	.0009	.0006
South Carolina	.0037	.0094	.0031	.0008
Georgia	.0025	.0076	.0013	.0008
Alabama	.0020	.0063	.0030	.0012
Mississippi	.0047	.0077	.0029	.0012
Arkansas	.0030	.0060	.0030	.0014
Louisiana	.0591	.0146	.0039	.0019
Texas	.0140	.0101	.0078	.0032

TABLE C16

Price Random Variance Estimates of Cotton; Selected States: Computed
for Granger, Greenback, Alliance, and Populist Periods

State	Random Variance 1867–1874	Random Variance 1874–1881	Random Variance 1883–1890	Random Variance 1890–1897
North Carolina	.00079	.00015	.00001	.00015
South Carolina	.00079	.00014	.00009	.00018
Georgia	.00079	.00012	.00002	.00013
Alabama	.00079	.00016	.00002	.00015
Mississippi	.00079	.00011	.00001	.00015
Arkansas	.00079	.00012	.00001	.00017
Louisiana	.00079	.00011	.00001	.00021
Texas	.00079	.00018	.00001	.00015

TABLE C17

Yield Random Variance Estimates of Wheat; Selected States: Computed
for Granger, Greenback, Alliance, and Populist Periods

State	Random Variance 1867–1874	Random Variance 1874–1881	Random Variance 1883–1890	Random Variance 1890–1897
Indiana	3.054	10.143	6.281	7.924
Illinois	.817	2.562	8.034	10.500
Wisconsin	2.214	4.567	2.339	2.424
Minnesota	2.714	8.020	3.746	8.531
Iowa	.696	7.297	.706	4.067
North Dakota	N.A.	N.A.	5.120	19.714
South Dakota	N.A.	N.A.	2.913	8.054
Nebraska	4.982	4.019	2.066	4.839
Kansas	2.676	5.001	9.074	14.500
North Carolina	1.138	.947	1.216	2.286
South Carolina	.786	1.123	.611	1.161
Georgia	1.781	1.216	.888	1.232
Alabama	.929	1.066	.494	.607
Mississippi	.304	2.067	.963	1.107
Arkansas	2.304	7.148	1.927	.924
Louisiana	N.A.	N.A.	N.A.	N.A.
Texas	3.589	7.437	2.613	9.388

TABLE C18

Yield Random Variance Estimates of Corn; Selected States: Computed
for Granger, Greenback, Alliance, and Populist Periods

State	Random Variance 1867–1874	Random Variance 1374–1881	Random Variance 1883–1890	Random Variance 1890–1897
Indiana	23.071	4.959	27.106	29.500
Illinois	55.736	22.429	37.686	40.554
Wisconsin	14.625	20.976	21.716	20.924
Minnesota	8.643	9.759	10.788	14.143
Iowa	17.786	10.509	31.709	80.982
North Dakota	N.A.	N.A.	21.623	6.357
South Dakota	N.A.	N.A.	19.268	91.875
Nebraska	37.893	28.276	42.505	95.321
Kansas	85.179	53.391	74.071	45.353
North Carolina	2.268	1.797	1.474	1.732
South Carolina	1.420	.714	.968	1.625
Georgia	1.987	1.088	.709	.500
Alabama	.674	.602	1.201	2.661
Mississippi	2.861	2.351	1.118	1.911
Arkansas	3.929	9.344	1.669	6.246
Louisiana	2.929	2.220	1.358	1.888
Texas	11.464	26.848	7.549	21.625

TABLE C19

Yield Random Variance Estimates of Oats; Selected States: Computed for
Granger, Greenback, Alliance, and Populist Periods

State	Random Variance 1867–1874	Random Variance 1874–1881	Random Variance 1883–1890	Random Variance 1890–1897
Indiana	7.250	8.643	8.998	22.138
Illinois	18.589	54.580	18.269	28.353
Wisconsin	14.710	11.246	12.768	6.424
Minnesota	6.929	20.924	8.159	19.603
Iowa	16.638	18.023	12.773	60.000
North Dakota	N.A.	N.A.	21.634	13.196
South Dakota	N.A.	N.A.	19.161	42.781
Nebraska	19.357	23.126	7.324	40.089
Kansas	24.571	17.321	6.286	38.996
North Carolina	4.317	1.483	1.103	2.625
South Carolina	5.143	1.960	1.710	3.429
Georgia	3.317	1.607	.714	2.089
Alabama	1.031	1.339	.161	6.960
Mississippi	1.893	.839	.603	2.196
Arkansas	3.554	10.911	2.071	2.821
Louisiana	1.674	1.071	2.482	5.746
Texas	1.071	6.143	5.357	20.214

TABLE C20

Yield Random Variance Estimates of Cotton; Selected States: Computed
for Granger, Greenback, Alliance, and Populist Periods

State	Random Variance 1867–1874	Random Variance 1874–1881	Random Variance 1883–1890	Random Variance 1890–1897
North Carolina	298.86	816.98	396.27	366.57
South Carolina	780.84	760.79	308.13	249.57
Georgia	635.71	380.00	97.21	118.71
Alabama	376.13	271.43	58.93	394.07
Mississippi	746.00	445.84	117.07	556.71
Arkansas	1429.13	2075.79	167.00	847.86
Louisiana	2154.14	1292.64	544.50	1756.29
Texas	649.98	1535.64	348.14	1462.00

TABLE C21

Income Random Variance Estimates for Wheat; Selected States: Computed
for Granger, Greenback, Alliance, and Populist Periods

State	Random Variance 1867-1874	Random Variance 1874-1881	Random Variance 1883-1890	Random Variance 1890-1897
Indiana	61904	16.492	.747	6.830
Illinois	91437	6.182	3.236	8.223
Wisconsin	71676	8.077	3.424	3.862
Minnesota	5.719	8.901	1.865	4.812
Iowa	5.248	6.143	2.052	2.623
North Dakota	N.A.	N.A.	8.262	8.847
South Dakota	N.A.	N.A.	4.349	4.602
Nebraska	4.330	5.774	5.220	3.475
Kansas	6.896	2.235	9.591	5.769
North Carolina	3.246	1.456	.668	1.947
South Carolina	3.842	3.808	.426	2.148
Georgia	7.798	3.311	.710	2.031
Alabama	1.808	3.578	.510	1.025
Mississippi	4.771	6.274	.838	1.813
Arkansas	10.014	8.040	1.435	.968
Louisiana	N.A.	N.A.	N.A.	N.A.
Texas	8.694	16.466	2.421	6.850

TABLE C22

Income Random Variance Estimates of Corn; Selected States: Computed
for Granger, Greenback, Alliance, Populist Periods

State	Random Variance 1867-1874	Random Variance 1874-1881	Random Variance 1883-1890	Random Variance 1890-1897
Indiana	2.926	3.564	1.435	2.523
Illinois	5.541	1.985	1.844	3.739
Wisconsin	13.052	7.765	3.687	1.115
Minnesota	15.220	3.978	1.567	.586
Iowa	6.846	3.760	1.280	1.345
North Dakota	N.A.	N.A.	2.254	1.988
South Dakota	N.A.	N.A.	.862	3.599
Nebraska	9.548	4.365	1.303	1.514
Kansas	4.149	1.404	.672	1.492
North Carolina	1.689	.715	.555	.727
South Carolina	1.253	.982	.420	1.471
Georgia	1.104	.729	.333	.655
Alabama	.995	1.793	.068	.956
Mississippi	3.563	1.949	.149	.463
Arkansas	3.689	2.082	.378	.998
Louisiana	3.517	.999	.475	1.398
Texas	10.231	1.969	1.683	2.871

TABLE C23

Income Random Variance Estimates of Oats; Selected States: Computed
for Granger, Greenback, Alliance, and Populist Periods

State	Random Variance 1867–1874	Random Variance 1874–1881	Random Variance 1883–1890	Random Variance 1890–1897
Indiana	1.530	.893	.511	2.616
Illinois	1.216	3.767	.741	3.975
Wisconsin	3.285	2.520	2.271	2.350
Minnesota	6.231	3.021	1.963	1.355
Iowa	4.927	2.796	3.376	2.288
North Dakota	N.A.	N.A.	2.310	1.885
South Dakota	N.A.	N.A.	1.249	.995
Nebraska	4.048	5.296	1.693	1.842
Kansas	3.199	2.779	2.054	1.406
North Carolina	2.980	.599	.140	.358
South Carolina	3.603	1.536	.525	1.006
Georgia	2.585	.720	.101	.654
Alabama	.362	1.462	.281	.113
Mississippi	1.096	1.824	.802	.291
Arkansas	1.864	4.759	.756	.626
Louisiana	10.301	2.123	1.392	1.752
Texas	6.838	7.347	1.413	2.688

TABLE C24

Income Random Variance Estimates of Cotton; Selected States: Computed
for Granger, Greenback, Alliance, and Populist Periods

State	Random Variance 1867–1874	Random Variance 1874–1881	Random Variance 1883–1890	Random Variance 1890–1897
North Carolina	8.460	21.754	4.602	4.455
South Carolina	17.711	22.246	1.848	6.914
Georgia	22.645	5.924	1.769	1.817
Alabama	8.392	6.674	.444	2.196
Mississippi	12.454	12.598	1.431	3.418
Arkansas	30.313	28.140	1.024	5.380
Louisiana	15.922	25.500	5.292	5.130
Texas	13.963	21.139	1.779	4.300

For Product Safety Concerns and Information please contact our EU
representative GPSR@taylorandfrancis.com Taylor & Francis Verlag GmbH,
Kaufingerstraße 24, 80331 München, Germany

Printed and bound by CPI Group (UK) Ltd, Croydon, CR0 4YY

01/05/2025

01858489-0001